Handheld Usability

Handheld Usability

Scott Weiss

Usable Products Company, New York

JOHN WILEY & SONS, LTD

National 01243 779777
International (+44) 1243 779777
e-mail (for orders and customer service enquiries): cs-books@wiley.co.uk
Visit our Home Page on: http://www.wileyeurope.com or http://www.wiley.com

Library of Congress Cataloguing-in-Publication Data

A Library of Congress catalogue record has been applied for

British Library Cataloguing in Publication Data

A catalogue record for this book is available from the British Library

ISBN 0 470 84446 9

Typeset in 10.5/13pt Sabon by Vision Typesetting, Manchester
Printed and bound in Italy by Rotolito Lombarda SpA, Milan, Italy
This book is printed on acid-free paper responsibly manufactured from sustainable forestry, in which at least two trees are planted for each one used for paper production.

Contents

Introduction

Overview of this Book

This book is a practical, hands-on guide to designing applications for handheld electronic devices. Handheld devices, in the context of this book, are computing and communication devices that are usable and useful while held in one's hands. These devices include email pagers, personal digital assistants (PDAs) and mobile telephone handsets. Laptop, palmtop and tablet computers with large displays are not included since they look and work like desktop computers.

I start by discussing the differences between handhelds and the desktop. Following this is an in-depth analysis of the components and features of handheld devices, including an overview of all current wireless communication standards, such as the wireless application protocol (WAP), i-mode, 802.11 (Wireless Fidelity, or Wi-Fi) and Bluetooth™.

The second half of the book covers the disciplines of information architecture, rapid prototyping and usability testing as they relate to handheld devices. Though it might appear these three disciplines are universal to desktop and handheld products, design for handheld products is decidedly distinct from design for desktop computers.

Application design for handheld devices is different from the design of desktop Web and software user interfaces. Handheld devices are used by people who are 'on the go', and the devices – as compared with desktop computers – have smaller displays, trickier input mechanisms, less memory and storage, and less-powerful operating systems. Understanding these challenges is the first step toward designing great products for handheld devices.

There are several good books about designing for the desktop Web, and even more about designing desktop software. There are, however, few books that cover handheld products at all and even fewer that focus on user interface design for this medium. This book will enable you to accomplish the following:

- understand the types of handheld devices and their differences
- design user interfaces for handheld devices
- design user interfaces for the wireless Web (WAP)
- prototype user interfaces for handheld devices
- conduct usability tests on prototypes and live handheld product applications

Who Should Read this Book?

This book is for designers and developers of applications for handheld devices. It assumes a working knowledge of the Web, mobile telephone handsets, and PDAs. It does not assume a working knowledge of technical specifications for the wireless Web, nor does it expect readers to be experienced in designing for the desktop Web or other platforms.

How Does One Design for a Handheld Product?

Understanding your audience is the first step. Knowing who they are and what they want to do helps you begin to understand their needs. People have instinctive ways of doing things. If someone is presented with a reasonable prototype and asked to do something with it, they will make an attempt. That attempt demonstrates their instinct and provides the designer with critical clues about how to design a product.

In this book, methods for 'usable design' are presented, including paper prototyping and usability testing. With product-specific user interface design guidelines and some practice, readers of this book will be able to design applications for any handheld device.

Timeliness vs. Timelessness

This book represents a 'moment in time', especially with regard to the discussion of handheld devices. The pace of technological innovation is extremely rapid. Devices come and go so quickly that some are barely noticed. Perhaps part of the reason that device designs are so temporary is that no valuable guides to designing for them currently exist. This book is my attempt to address that problem.

Since the pace of design and development in the handheld arena has been so fast, many companies have reinvented the wheel, sometimes more than once. A glaring example of this problem is the lack of a standard layout for mobile telephone handset controls. The 'Back' button is the most popular control in desktop Web browsing, but only a small fraction of the dozens of Internet-enabled handsets has one. Using the Web without the 'Back' button is like using a word processor without being able to 'Undo'.

The result of these inconsistencies is a general lack of usability on most handheld devices. Learning about the different devices and their capabilities will enable you to design applications that are easy to use by themselves as well as with other applications found on the same platform.

Although technology and industrial design will march forward, this book's chapters on information architecture, prototyping, and usability testing will persevere. The principles and methods described here are timeless in scope. No book has been written about the design of handheld applications, a new discipline.

Usability and its Value

Usability is most easily defined as 'ease of use'. Usability testing is the objective study of a product's ease of use by watching people while they attempt to complete tasks. Whenever the test participants encounter difficulty with a particular task, there is a usability problem. Users who encounter difficulty are not 'stupid' – bad design is.

Planning usability testing as part of the design and development process makes economic good sense. Product designs that are evolved through usability evaluation prevent customer support calls and increase customer satisfaction. Happy users become loyal customers, and with so many handheld devices to choose from, usability can be a powerful distinguishing factor.

Chapter Breakdown

Each chapter in this book is intended to stand on its own. In other words, if you feel that you know enough about information architecture, then jump right to the chapter on usability testing without hesitation. If you find that you are in 'over your head', you can always step back to the chapters on information architecture.

The book is structured as follows:

- Chapter 1, Handheld vs. Desktop. This compares and contrasts handheld devices and the desktop computing platform.
- Chapter 2, Handheld Devices. This breaks handheld devices down into the three types – pagers, PDAs, and phones – and describes wireless networking standards, WAP, and i-mode.
- Chapter 3, Information Architecture: Process. This teaches the discipline of information architecture and all of its steps: audience definition, scenario development, flow charting, application mapping, and page mapping.
- Chapter 4, Information Architecture: Practice. This gives a full breakdown of software user interface constructs for handheld devices and when and how to include them in designs; common user interface types are described, and specific sections are included for WAP, Palm OS®, Windows CE/Pocket PC, RIM OS, and Motorola Wisdom™ platforms.
- Chapter 5, Prototyping. This presents the different types of prototyping and teaches paper prototyping and online prototyping methods.
- Chapter 6, Usability Testing. This describes usability testing and all of the issues unique to testing handheld devices; it provides detailed descriptions of the process and execution of formal (lab) testing and testing rapid prototypes.

In addition to the chapters, there is a rich set of appendices to help you in your handheld device design, prototyping, and usability testing efforts:

- Appendix A, Handheld History. This provides a timeline about technologies leading to handheld devices.
- Appendix B, Paper Prototyping Applications for the Palm OS®. This gives a step-by-step, illustrated example of how to create a paper prototype, using the built-in Address Book found on Palm OS® handhelds.

- Appendix C, Sprint PCS/NeoPoint 1000 Usability Study. These give the full usability study conducted by Usable Products Company in Winter 2000 of the Sprint PCS Wireless Web using the NeoPoint 1000 handset.
- Appendix D, Glossary. Here, hundreds of terms about handheld devices, wireless networking, and mobile telecommunications are defined.

Conclusion

This book shares with you methods of design, prototyping, usability testing, and documentation that I have worked to develop for 12 years. I have looked for many of these methods in print and not found them, despite some books' promising titles. It is my sincere hope that *Handheld Usability* will help to bring about more enjoyable and easy to use applications for handheld devices.

Author's Note

I have been fascinated by handheld computing devices since the introduction of the TRS-80 Pocket Computer Model 100 in 1983. I was amazed that a computer could be carried around, since most of the computers available at that time were bulky table-top models. The TRS-80 Model 100 had 24K of memory, a full QWERTY keyboard, a 240 × 64 pixel monochrome bitmapped display, and weighed about 3 pounds. Radio Shack was prescient enough to include a bar-code reader and a modem, perhaps anticipating the use of handheld computers as information appliances. Model 100s are in use even today. You can read a timeline tracing the development of handheld devices in the Appendix A, Handheld History.

The TRS-80 Model 100 of 1983 is a far cry from today's hottest Palm OS communicators, which are mobile telephone headsets with the Palm OS platform built in. The Handspring Treo, with 16 megabytes of memory, a QWERTY keypad, and 160 × 160 pixel gray-scale bitmapped display, weighs only 5.4 ounces (154 gs). It is also small enough to fit in a shirt pocket.

In exploring user interface design for these devices, I researched the available design guidelines for applications on all of the available handheld device platforms. Finding only limited product-specific guidelines and technical programming information, the opportunity to research and write about designing for handhelds presented itself. I am pleased to offer you this book as the result.

Scott Weiss

About the Author

Scott Weiss (sweiss@usableproducts.com) is the principal of Usable Products Company, the industry leader in the field of handheld device information architecture and usability testing. Usable Products has helped Sprint, Dun & Bradstreet, Intel, Chase, GlaxoSmithKline, and many others with their information architecture and usability needs since 1996.

Scott's design work on Apple's Macintosh System 7 and Microsoft's Windows 95 can be seen on more than 90 percent of computer desktops worldwide. In a career that has spanned desktop software and the Web, Scott has consulted, taught, and lectured extensively on usability and information architecture for handheld devices, desktop software, and the Web.

Scott received a Bachelor of Science degree in Engineering Human Interface Design from Stanford University. He lives in New York City with his bearded collie, Midge.

Acknowledgements

This book was a project not just for me, but for my employees, consultants, friends and colleagues, all of whom contributed by providing feedback and moral support throughout the writing and editing process. I started this project soon after returning from the 2001 annual meeting of the Association of Computing Machinery's Special Interest Group on Computer–Human Interaction (ACM SIG-CHI).

First and foremost, I would like to thank Hira Murtaza for her excellent illustrations. Her work brings beauty and clarity to this material. Jerry Weinstein was an invaluable muse and editorial assistant. Robert Eisenhauer and Richard Martin, who work with me at Usable Products, were invaluable in their input and management of much of the editing process. Both Rich and Robert managed the extensive permissions necessary for including device and software photographs, and Rich conducted extensive research for *Handheld History*. I would also like to thank Karen Mosman and Jill Jeffries, my editors at Wiley, for their encouragement.

Kent Sullivan of Microsoft was kind to make referrals to members of the Pocket PC team. Alastair France of Openwave demonstrated the new Openwave browser at a meeting in London, far from his work and family. His technical prowess rescued a meeting of the Wireless Roundtable. Mark Taguchi of Openwave met with me to discuss the GSM's M-Services Guidelines. Erin Davis set up an interview with Calin Pacurariu, the Handspring Treo Product Manager. Calin Pacurariu was kind to take the time to meet with me about the Handspring Treo. Richard Weeks provided insight, information, and referrals. Eli Katz helped edit the Introduction.

I would like to thank Maria Sääksjärvi for sending content through the mail, and Clifford Nass and Ing-Marie Jonson for their contributions. Eddie Gomez provided his insights into usability of Windows CE-based tablet hardware and software. Kylie Trevitt prepared an excellent history of Openwave. Joshua

Seiden provided helpful links for Chapter 4 as well as for Appendix A on handheld history. Robert Moritz, of Sprint PCS, answered many technical questions. Christie Hardin of the Motorola Media Intelligence Center helped gather information for Appendix A, as well as helping with equipment evaluation requests and fulfillment. Courtney Flaherty helped with information about and an evaluation of the RIM Wireless Handhelds™. Scott Jenson answered questions about Symbian OS. Joan Schnorbus answered questions about UPS. Andrew Hicks helped resolve confusion about the Ericsson R380 communicator. Adam Tow answered questions about the Apple Newton. Del Penny sent 35 mm slides of the Apple II and Newton 2100. And Midge, my faithful companion, kept me company all the while.

Handheld vs. Desktop

This book is about designing applications for handheld electronic devices, specifically mobile telephone handsets, pagers and personal digital assistants (PDAs). To understand handhelds, one must first understand their place on the continuum of personal computing devices – and also their place within portable consumer electronics. Finally, one must understand the three key components of handheld electronic devices – and how they overlap. For a comprehensive history of technologies leading to handheld devices, see the Handheld History in Appendix A.

This chapter compares handheld devices with their closest cousins – desktop computers. Handheld devices are very similar to desktop computers in that they each involve computation, information management, and communication. The desktop computer, however, is far from portable, but that difference is only one of several, which this chapter quantifies.

Definition of Desktop Computers

No comparison between 'handheld' and 'desktop' would be complete without a definition of what constitutes a desktop computer. Every modern desktop computer has the following components:

- CPU (central processing unit), the 'brain'
- display

- keyboard
- pointing device, typically a mouse
- cables to connect the components to each other, to power, and to a network

Some desktop computers combine components, such as the Apple iMac, which combines the CPU and display, but most desktop computers require five separate components to function effectively.

Definition of Handheld Devices

Handheld devices are extremely portable, self-contained information management and communication devices. A candidate must pass three tests to be considered a handheld device:

- it must operate without cables, except temporarily (recharging, synchronizing with a desktop)
- it must be easily used while in one's hands, not resting on a table
- it must allow the addition of applications *or* support Internet connectivity [wireless application protocol (WAP), i-mode, or email]

I developed these tests as a way to frame this book. The tests eliminate laptop computers and noncomputer portable consumer electronics devices such as MP3 players. While this book specifically addresses only handheld devices, the principles presented are applicable to all devices that satisfy one or more of the tests above.

The Personal Computing Continuum

Nomenclature is a particular challenge in any new area of technology. For this book, the term 'handheld' came to me after much thought and discussion. In earlier iterations, I considered 'wireless' and 'mobile'. However, each of those terms alone captures only part of the meaning of 'handheld'. The same consideration went into many other term selections, such as 'palmtop', 'pager', and 'communicator'. In this book, each term is defined when it is introduced, and there is a glossary for your convenience, at the end of the book.

In the illustration below, four different types of personal computing devices overlap to show how the categories are not discrete. Some devices are difficult to quantify, such as ultra-small all-in-one desktop units, which look and act more like laptops. However, most devices comfortably fit into one of the categories shown in Illustration 1.1.

Size decreases to the right as portability increases. Desktop computers, at the far left, are stationary devices, tethered for both power and connectivity. Laptop computers enable mobility, but they are heavy and must be used on a table. They work best when connected to a power source, and most laptop users travel with telephone or Ethernet cables to connect to corporate networks and the Internet. Palmtops look like laptops, but they are significantly smaller, often fitting into a pocket or purse. They typically run off batteries, but need to be recharged frequently. At the far right are handhelds, which function best while held in the hand.

All of these devices can connect to the Internet, although some require modems or expansion cards to do so.

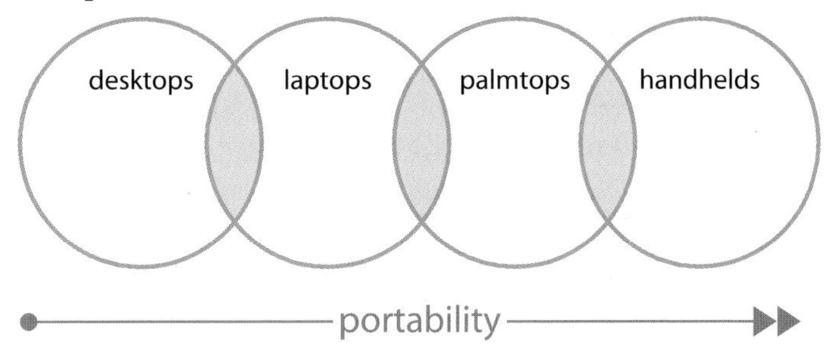

Illustration 1.1 The personal computing continuum

Handheld Devices

Illustration 1.2 shows how the category of handheld devices is further segmented. At the edges are the three device types covered by this book. In the center, the overlap of all three types, is communicators, which offer all of the features of handheld devices: voice, computing, and Internet connectivity. In the next chapter, 'Handheld Devices', I describe, compare, and contrast each of these handheld device types in detail. For now, let's continue with the comparison of handheld and desktop devices.

Portable Consumer Electronic Devices

Illustration 1.3 places different types of devices in the domain of portable

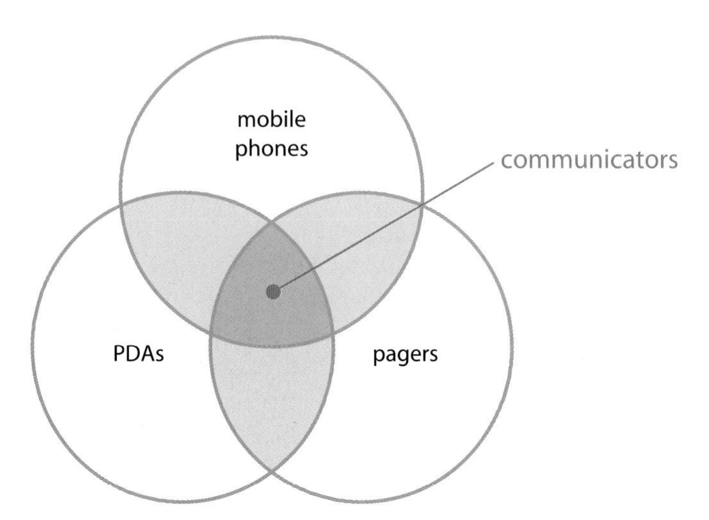

Illustration 1.2 Handheld devices

consumer electronics. While this book covers design for handheld devices, it excludes most portable consumer electronics devices, such as handheld electronic games, calculators, digital cameras, and MP3 players. Most of those devices lack the ability to add applications and lack wireless Internet connectivity, so they fall outside the purview of this book. In their capacity of allowing applications to be added, gaming machines are covered by this book, but not with respect to gaming user interfaces. Game design is very different from other user interfaces and is well covered by other books, such as *Game Design: The Art & Business of Creating Games* (2001) by Bob Bates and Andre Lomothe, and *Game Design: Theory and Practice* (2001), by Richard Rouse and Mark Louis Rybczyk.

Handheld vs. Desktop

The desktop platform is stationary, while the handheld platform is extremely portable. In-between are portable devices that require a table to be operated effectively, such as laptop and palmtop computers. Table 1.1 sums up handheld vs. desktop devices.

As you can see from the table, handhelds are smaller, lighter, and less 'capable' than desktop computers. However, they offer portability and instant access to time-critical information as tradeoffs. Designs for handheld devices must capitalize on their advantages and accommodate their weaknesses.

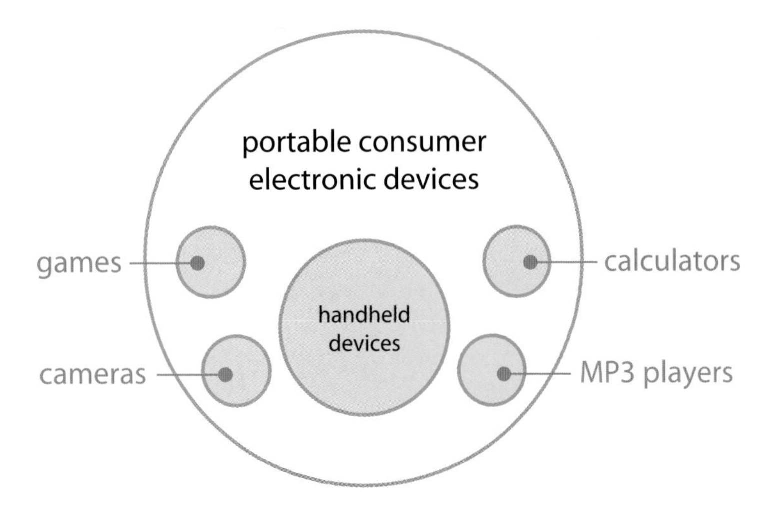

Illustration 1.3 Types of portable consumer electronic devices

Table 1.1 Summary of platform characteristics

	Handheld	Desktop
Reasons for Use	On-the-go lookup and entry of information. Quick communication, such as instant messaging and/or voice.	Lengthy information processing tasks. Web browsing and email.
Form Factor	Less than 10" × 12". Can be used standing up without a table, typically fitting in a shirt pocket.	Requires a table; best used while seated. Sometimes portable.
Mobility	Can be easily used while on the go; has great battery life	Requires a power cord, or frequent recharges
Connectivity	Slow and unreliable, but improving.	Fast and reliable.
Input	Challenged. Stylus, keypad, mini-keyboard, roller wheel.	Full keyboard and mouse.
Display Size	240 × 320 pixels or less.	640 × 480 pixels or more.
Memory	Up to 16 megabytes.	Up to 1 gigabyte.
Storage	Sometimes none. Removable Memory Stick® media available on some models with up to 128 megabytes of storage.	Diskettes, CDRW, DVD RAM rewritable media, as well as hard disks with 80 gigabytes or more of storage.

Understanding the differences between the two platforms will enable you to design excellent applications for each.

Reasons for Use

Handheld devices and desktop computers are used for very different functions, although both have information management and communication features. Both devices often can synchronize their data as well. Applications designed for one type of device require rethinking in order to be useful on another device type.

Desktop computers are used for lengthy data processing applications, such as word processing, financial analysis, and computer-aided drawing. Handheld computers, since they are small and lightweight, tend to be used as personal information managers, portable compressed document viewers, and in-the-field data entry devices.

Desktop computer users remain in their seats in front of their displays for hours at a time. Handheld device users may glance at their schedule to learn where to go for their next meeting or to look up a telephone number just before placing a call – sometimes on the same device, sometimes not. Mobile telephone handsets and communicators are used for voice communication but offer messaging and/or Internet access as well.

The theme that is emerging from this discussion is *access to information*. Handhelds require fast access to information, while desktop computers require comprehensive access. For handhelds, the tradeoff for speed and portability is completeness.

Form Factor

Form factor is a device's size–shape relationship. Handheld devices are easily held in one hand and operated with the other. Handheld devices with keyboards are typically held with both hands and operated with both thumbs. Desktop computers come in many form factors, such as the under-monitor, tower, compact, and all-in-one.

My true test for a handheld is whether it requires a table to be operated effectively. The Psion Revo Plus device, shown in Illustration 1.4, fails this test. Although it is very small and when closed would fit in a shirt pocket, it works best while resting on a table. These devices are classified as 'palmtops'. Palmtops more closely match desktop computers than they do handhelds, which is why they are not specifically addressed by this book. In order to design for

Illustration 1.4 Psion Revo Plus

palmtops most effectively, refer to books on desktop and Web application design as well as the sections in this book on PDA design, such as in Chapter 2, on Handheld Devices, and in Chapter 4, on Information Architecture.

The Fujitsu PenCentra 200, shown in Illustration 1.5 also is not a handheld simply because of its size.

This device sports a 640 × 480 pixel display, is 8.9" × 6.5" × 1.3", and weighs 2 pounds. Its large touch screen display makes it work much more like a desktop computer than a handheld device. As you can see in the photograph, the display is of a Windows desktop, so designs for this device should follow user interface guidelines for Windows desktop software.

Illustration 1.5 Fujitsu PenCentra 200. Reproduced by permission of Fujitsu PC Corporation

Mobility

Desktop computers are obviously not 'mobile'. They are meant to be stationary. Handheld devices define *mobility*. They are small, lightweight, and very portable. Defining the traits that make a device 'mobile' requires a deeper look into quantifying the concept.

Mobility is determined by the following factors:

- combined weight of all necessary components, such as CPU, display, keyboard, and mouse
- number of required components and how they are configured
- cables needed for operation, such as power, networking, and component configuration
- size of fully configured system in height, width, and depth
- furniture necessary for system configuration

Weight

Weight of desktop computers is measured in pounds, but usually just for shipping purposes. The weight of a handheld device is calculated in ounces or grams. Typically, handheld devices weigh less than 8 ounces.

Components

Desktop computers are usually meant to rest on a table and require displays, input devices (keyboards and mice), and cabling to be usable at all. Handheld devices are often all-in-one units, but they frequently require a stylus input device, which is easily stowed within the unit. Handheld devices sometimes have attachments such as headsets, power connections, modem cables, etc., but they are usable as independent units most of the time. Some add-on components can be snapped directly onto a device, while others travel separately. The key is that all components for handheld devices must be portable or stowable. The exception is a cradle that is connected to a desktop computer for synchronization and sometimes recharging.

Cables

Desktop computers require many cables, to connect components. Handheld devices typically require only a power cable, when they are recharging. Cables may also be used for connectivity, headsets, or, infrequently, for connectivity.

Size

Desktop computers take up more than half of a typical desk surface. Handheld devices fit in a shirt pocket.

Furniture

Desktop computers require surfaces on which to be placed. Handheld devices require only one or two hands, except when charging, synchronization, or when not in use. Mobility is critical to handheld devices, and is part of the equation of compromise the consumer must make when purchasing one. Smaller, lighter handhelds might be lacking in battery life or have a smaller display. Mobility of desktop computers is not generally an issue, but becomes critical for laptops and palmtops.

Connectivity

Desktop computers and handheld devices feature connectivity, to each other and/or to the Internet. The fundamental difference between them is that desktop computers are typically always connected, while handheld devices make and break their connections based on need. Connectivity requires power, and battery life is at a premium, so even 'always-on' connections deactivate after being idle for a predetermined period.

Desktop computers rely on connectivity for transferring large volumes of data, whereas handheld devices transfer small amounts of data. The main reason for this difference is that handheld devices have much less available storage, but slow transmission speeds also are a significant factor. When available, handheld devices are more likely to transfer streams of data, rather than receive data that will be stored on the device. Video and audio are frequently streamed to the desktop today, although many desktop users store files so they can replay the media.

Voice

Voice connectivity is the primary means of connectivity for mobile telephone handsets and communicators and is usually not available on other handheld devices. Voice connectivity is rarely available on desktop computers, except when part of videoconferencing.

Beaming

Handheld devices can transfer bursts of information to other handheld devices via **beaming**, which today relies on infrared, line-of-sight transfer. Desktop computers do not transfer data in that manner at all but instead rely on more sophisticated networking protocols that provide additional security and reliability. Beaming is like using a remote control to change the channel on a television, whereas desktop computer networking is more like the built-in control panel on the same television. The remote control transfers a burst of data in a somewhat unreliable manner, whereas the control panel is hard-wired and much more reliable and secure.

Synchronization

Synchronization is a unique type of connectivity in that it is not in real time. Handheld devices and desktop computers synchronize calendars, address books, movie listings, email, and other content. Synchronization is typically performed via cable, although newer devices feature wireless synchronization capabilities. Some synchronization protocols work via modem and/or the Internet, but by far most synchronization is done via direct cable connection.

Desktop computers do not synchronize with each other, and handheld devices do not synchronize with each other either. Synchronization is strictly between desktop computers and handheld devices.

One of the most successful applications designed specifically for synchronization is Vindigo™. Vindigo™ is an entertainment resource with updated movie, restaurant, and shopping information on a city-by-city basis. Vindigo™ is available with synhronization on PDAs and via WAP for mobile telephone handsets. For PDAs, Vindigo™ gathers updated content through the desktop computer's Internet connection during the synchronization process. Vindigo™ and applications like it make the best use of speed for content that needs only to be updated once each day or less. The desktop Web is much faster than the wireless Web, and synchronization is a fast way to transfer large amounts of data.

Wired vs. wireless

Most desktop computer networks are **wired**, meaning that they require physical cables to function. In addition to cabling, desktop computer networks require specialized networking hardware, called hubs and routers, to amplify and direct the networking signals. For desktop computers, wireless networking has the benefit of saving time and resources, since extensive cabling is not required. However, hubs and routers are still required, which *are* wired into the Internet. Wireless desktop networking is generally not designed to make the desktop more mobile, which is the main benefit of wireless networking for handheld devices.

Handheld devices can be networked to local area networks (LANs) and/or to the Internet. For LAN connectivity, 802.11 or Bluetooth™ are used. Both are technologies that offer fast, reliable connectivity. When a LAN with Internet access is not available, handheld devices need to rely on cellular connectivity, which is slower and less reliable, although it is available in most geographical areas. Networking standards are discussed in detail in the section on connectivity (pages 52–56) in the next chapter, on Handheld Devices.

Satisfyingly fast (640 kbps) vs. intolerably slow (9.6 kbps)

Desktop computers enjoy extremely fast networking, with DSL and cable modems offering at least 640 kbps. The user experience of 640 kbps is that a web page rich with graphics and animated imagery will appear within a few seconds on a recent model desktop computer. However, most handheld devices connect to the Internet at extremely slow speeds, such as 9.6 kbps.

Latency, the amount of time one has to wait before a data transfer starts is another consideration. Most desktop networking configurations have imperceptible latency. However, today's handheld device Internet connectivity has high latency, sometimes up to 10 seconds for an initial connection. The reason for the high latency in handheld devices is the nature of the connection. Most handheld devices rely on circuit-switched data, which requires a dedicated connection to be made before data can be transferred. All desktop computers today use packet-switched data, which features 'always-on' connectivity.

For non-Internet connectivity, handheld devices can connect at satisfying speeds using either 802.11 or Bluetooth™. These and other wireless networking protocols are described in detail in the next chapter, on Handheld Devices (pages 54–56).

Stable vs. Unstable

Handheld devices have unstable Internet connectivity, since they rely on an overburdened, problematic networking infrastructure. Desktop computers rely on networking infrastructure that has been continually upgraded for many years. While desktop Internet access may not always be satisfactory, it is nearly always reliable.

Input

Entering data on a desktop computer is easier and more standardized than on a handheld device. It is far easier to enter text on a full-size keyboard than on a 12-key keypad, for example. Clicking and dragging with a stylus on a tiny touch screen can never match the ease with which one can use a mouse with its corresponding pointer on a large display.

Desktop computers rely on the keyboard and mouse as their primary means of input. Handheld devices feature touch screens and/or keypads. In addition to the primary means of input, accelerators are often available, such as roller wheels, rocker controls, labeled buttons, and softkeys. All of these input methods are described in the next chapter, on Handheld Devices (see the section on Data Input, pages 43–49).

Output and Display

Desktop computers rely on large displays of at least 640 × 480 pixels. Most desktop computers today are capable of display resolutions of 1024 × 768 pixels or even denser resolutions. Display sizes range from 15 inches to 21 inches, measured diagonally. Handheld devices, on the other hand, top out at 320 × 320 pixels, and that high resolution is available only on a small fraction of the largest handheld devices. Some handhelds are character-only and can only display 4 lines of 16 characters. Color display is ubiquitous on desktop computers but is a rare option on handheld devices. When color is available, it often is a poor representation, since handheld devices have little memory, touch screen coatings, and must be usable in direct sunlight. With little memory, far fewer discreet colors are available, and the requirement that handheld devices have touch screens and that they must be usable in drastically different lighting conditions means that some visibility is sacrificed.

Printing is another challenge for handheld devices. Desktop computers connect to printers extremely well, but handheld devices typically offer no printer connectivity. There have been some forays into this area, but the technology has not been perfected.

Desktop computers support all types of rich media, including audio and video. Some of the highest-end handhelds can play music and show video content. However, the memory and display limitations of handhelds make multimedia features more of the promise of tomorrow than the reality of today.

Memory and Storage

With limited memory and little or no storage capabilities, handheld devices are better suited to browsing data available through a network than storing it locally. RAM (random access memory), the primary 'thinking' power of a computing device, is more constrained on handheld devices than on desktop computers. Today's desktops can hold 1 gigabyte or more, whereas handhelds are often limited to 16 megabytes or less. In real terms, a desktop computer can run a CAD (computer-aided design) application that can design a skyscraper, whereas a handheld device might only be able to display a vastly simplified version of the same drawing – and would probably not be able to modify it.

Storage is another matter. Desktop computers have vast amounts of disk storage available and can write to removable media such as diskettes, CD ROMs, or DVD RAM. Whereas diskettes store only 1.4 megabytes of data, DVD RAMs can hold gigabytes of data. Most handhelds do not support removable media at all, and the ones that do support the Sony Memory Stick® or comparable products that are presently limited to 128 megabytes. Handheld device storage is a pressing problem that will be addressed in coming evolutions of handheld hardware. Digital cameras are most likely to lead the way in storage solutions.

Design Differences

Desktops rely on different regions of the screen, layered windows, and multiple types of windows. Handheld devices can rely on some of these features some of the time, but, on small displays, additional user interface controls tend to add clutter rather than increase usability.

Boot time vs. instant on

Desktop computers require time to 'start up' from when their power is activated. Sometimes startup sequences can take several minutes. Handheld devices start up almost instantly. This fundamental difference requires many sacrifices in handheld device application design, as the quick startup feature is no miracle. Operating systems on desktop computers are stored on disk to make them easy to update and customize. Operating systems on handheld devices are stored in read only memory (ROM), which is less flexible than disk but faster and less power hungry. Handheld devices' operating systems are also much simpler than those of desktop computers, another sacrifice made to speed startup and keep hardware requirements slight.

Metaphor discussion: desktop

The desktop is a place where one stores documents, folders, office tools, and office supplies. Desktop computers have relied on the 'desktop metaphor' since 1981, with the introduction of the Xerox Star. Handheld devices, since they typically have little storage capacity, have little need for a 'desktop'. However, as handhelds support increased storage capacity, additional desktop elements will appear.

File system

Desktop computers make use of hierarchical file systems that store documents in folders that can themselves be placed in folders. Windows CE and Symbian OS feature file system support, but the Palm OS® platform, pager operating systems, and WAP do not. On handhelds without file system support, users must start with applications in order to locate data, and data available to one application is not available to another.

In early PDA designs, leaving out rich file system functionality was a way to keep the devices lean, small, and inexpensive. The absence of file management also proved to be a boon to ease of use. However, as memory and processor power have increased, the lack of rich file system support has begun to hamper the Palm OS platform and WAP. Products such as Documents to Go for Palm™ handhelds offer a file repository, but, unless they are rewritten, other applications for the Palm OS platform cannot take advantage of the features that a file system affords. Newer WAP browsers allow users to download and manage ring tones, graphics, and games and so have limited file system ability.

Multiple applications

Desktop operating environment offers users the opportunity to run multiple applications simultaneously and to have multiple documents open within each application. The ability to run multiple applications requires processor power, memory, and power. Some handheld environments do not offer this capability, and for good reason. Dedicating screen real estate to a bar listing the running applications is a luxury that handhelds can ill afford. The workaround is that most handhelds have an **application repository** where all the main application icons reside. The application repository is similar to a website's home page in that it is the place from which most top-level functionality can be accessed and is one that users can easily find on their own. The application repository differs from a website's home page in that it is rigidly designed and lives locally on the device.

Windows CE does support multiple simultaneously running applications. It allocates space at the top of the display so that users can switch between active applications. The Palm OS platform does not offer this functionality; users must use the 'home' icon to return to the application repository in order to switch to a different application. When they do, the first application is suspended.

Most handheld devices that do not support multiple simultaneous running applications retain state as a workaround. 'State' could be the data entered by the user, the last accessed information, or the results of the last calculation. State is saved from session to session, even when the device is powered off. When returning to the calculator on most handhelds, a user will be pleased to find that the running total he or she left behind is still active. State is remembered on a case-by-case basis. Address book applications tend to return the user to the directory, regardless of from where they left the application. This design choice relies on the likelihood that the next time the user enters the address book, he or she is likely to want a *different* person's contact information. 'Forgetting' the state in this case saves the user a step.

Auto save

Handheld devices always save data automatically. No deliberate 'save' action is required. Some desktop applications have explored this strategy, but most have opted against it. An exception is calendar applications, which have no 'save' option. One's calendar is a special type of file that is frequently updated, and requiring the user deliberately to save changes might result in a loss of data. More often than not, all calendar changes are desired, and an explicit 'save' requirement would be irritating.

Internet use differences

The desktop Web is open and vast. Millions of websites are available to nearly every desktop computer user. Most handheld devices do not have direct access to the desktop Web, and when they do they must rely on early generation hypertext markup language (HTML) Web browsers that compress or 'clip' graphics and have no support for Web technologies such as frames or Java-script.

The wireless Web is far less interesting than the desktop web because of all the limitations of handheld devices: less memory, problematic data input, small displays, etc. The wireless Web includes WAP sites and a small number of HTML sites designed specifically for small displays. WAP, which is available from mobile telephone handsets and some other handheld devices, is a variant of HTML, the language used to encode desktop websites. Many content providers are bypassing writing specialized HTML for handheld devices and are instead writing custom Internet-enabled applications. MapQuest (www.mapquest.com), an excellent example of this genre, is a tool that maps locations and provides driving directions to its users. Custom Internet-enabled applications take advantage of the rich user interface features of the handheld device and the connectivity of the Internet, without the limitations of handheld device Web browsers.

Surf vs. hunt Desktop Web users 'surf', using the Internet as an entertainment medium. The fast processors and high-speed data connections that are the norm today have made surfing enjoyable. However, handheld devices have slow processors and slow, faulty Internet connections. Combined with small displays and difficulty entering website addresses, the result is that wireless Web users 'hunt' for their information. They go to the Web with a purpose in mind and do not get easily side-tracked.

Unlimited use vs. cost per minute or per byte Desktop computer users in the USA and Europe have come to enjoy fast, unlimited Internet access for a fixed monthly fee. However, Internet access for most handheld devices comes at a dear per-minute price, comparable to that of voice calls. Most carriers also charge a monthly fee simply to provide access to the wireless Web, in addition to per-minute charges.

GPRS (general packet radio service), already available in Japan, is swiftly becoming available in the USA and Europe. This service changes the cost structure for data transfer from per-minute to per-byte (or per-megabyte). GPRS and other always-on, cost-per-byte solutions are likely to increase the popularity of wireless data products for consumers, who do not see the value in paying high per-minute prices for slow, unreliable data transfer.

Open landscape vs. walled garden Connecting to the wireless Web from a mobile telephone handset is very different from connecting from a desktop computer. Mobile handset connections are initiated by opening a browser, but what launches is a listing of website categories, much like a desktop Web portal. The 'Open URL' option is usually hidden at the bottom of the list, or even behind a couple of configuration pages. For this reason, the wireless Web is often called a 'walled garden', since users can get out only with difficulty, if at all. The desktop Web lacks a 'walled garden' organization of Web content, so I refer to it as an 'open landscape'.

The situation for other handheld devices is better. PDAs and pagers with Web access provide portals for convenience, but they also make it fairly easy for users to break out to the desktop Web, where some sites work well on small displays, and there are others that are specifically designed and formatted for handheld devices.

Bookmarking Since there millions of websites, bookmarking is an extremely popular feature. Bookmarks enable users to access frequently used websites with ease. On desktop computers, bookmarks are simple to create and access. Bookmarks are also easy to store on PDAs and pagers with Web access. On most mobile telephone handsets running WAP, however, bookmarks are difficult to store. New guidelines for mobile telephone handset user interfaces include bookmarking capabilities, however, so this limitation will disappear within the next couple of years.

Privacy and security (phones are private and secure) Desktop computers are often used by more than one person, and so privacy and security are significant concerns. Handheld devices are generally used only by their owner and so they are private and secure by nature. These devices are also very personal and are easy to personalize as well. Passwords are often cached on handheld devices as a result, providing convenience to users who might be irritated by having to enter a password each time they check email.

However, the small size of handheld devices makes them easy to lose, and their value makes them theft targets. Most handheld devices have password support. A typical implementation locks the device until the password is entered. If the password is forgotten, the user has the option of resetting the device and clearing all of its password-protected data.

Design redux

Handheld devices were designed as portable information management systems, reliant on desktop computers for data transfer. Since their debut, handheld devices have acquired wireless Internet access, making them more independent, but not fully so. Synchronization between devices remains a prevalent user interface.

The overall design aesthetic of handheld devices is that they are desktop computers' 'little brothers'. Handhelds offer a carefully selected subset of the features of desktop computers. Recently, handhelds have become useful in the field, where they can be used to capture and look up data in a professional setting. Design of these information applications requires different approaches than for desktop software, as a result of their strengths and their constraints.

Usability Evaluation Differences

Usability testing is a universal discipline, involving observing test participants as they attempt to complete tasks using software and/or hardware. Desktop usability testing has become standardized for formal research in a lab setting. Handheld devices, since they are small and mobile, pose unique challenges to direct observation, testing strategies, and test scenario development.

Lab setups

Desktop usability tests are often video recorded. The video recordings include both the desktop computer image and an inset of the user's face. The desktop computer video is converted to video format with a hardware device called a **scan converter**. The signal from the scan converter is combined with the signal from the video camera trained on the user's face, with another hardware device, called a **video mixer**. The signal from the video mixer is recorded on videotape.

Handheld devices do not support scan conversion, and so it is a challenge to record the display. Training a second video camera on the display is effective only if the device is mounted, which makes the device immobile, in direct contrast to how handheld devices are used. Attaching a camera to the device itself has other problems, such as making the device much heavier and adding cables where none are customary. Long tests also require constant power to the device, adding more cabling to the setup.

Some researchers have attempted to have test participants wear video camera 'hats', but every time the participant moves his or her head the image moves with it, making viewing the video recordings very challenging. I feel that the best compromise is to mount the device so that a video camera can be trained on the display. The mounting that I use is a tabletop tripod, to which we attach handheld devices with Velcro®.

Less formal

For less formal usability research, the method of research can be more like the medium. For desktop studies, research can be in an office setting, while, for handheld device research, the setting can be in a conference room, at a desk, while walking, etc.

Usability testing of rapid prototypes, which can be constructed from paper or on-screen, are very similar for desktop software and handheld devices. The settings are the same, and the methodologies are the same. In the Prototyping and Usability Testing chapters. I present a complete solution for creating prototypes and testing them (Chapters 5 and 6).

Conclusions

The differences between desktop computers and handheld devices require different design strategies. Handheld devices, with their increased mobility, limited memory and processing power, and small display sizes offer unique challenges – and opportunities – for design.

The desktop computer is not going away, and neither is its relationship to handheld devices. Desktop computers have changed a great deal since the introduction of the Apple II in 1977, primarily with the introduction of windowing environments and direct manipulation using a mouse. Handheld devices, which were introduced in 1993 with the Apple Newton, have already evolved significantly and are poised to change a great deal more with the advent of always-on wireless Internet connectivity. Most handheld application design today relies on transfer of files from the desktop, and synchronization of content, such as address books and calendars, between both devices. Tomorrow's handhelds will be less reliant on desktops but will interact with them more freely, and wirelessly, too.

Summary

Desktop computers are made up of several components, are large, heavy, require furniture, and must be attached to a constant power source. They are useful for stationary workers for intensive information management applications. They also provide speedy Web access, have large amounts of memory and storage, and feature large displays. Handheld devices are small and lightweight, often made up of a single component. They are useful to people 'on the go', for quick access to personal data and lookups of information using the wireless Web. The sacrifices for speed and small size are less memory and little or no storage, small displays, and slow, unreliable access to the Internet.

Since the devices are very different, design strategies for them are different as well. Desktop computers take more time to start up but offer massive storage, sophisticated file management capabilities, allow multiple software applications to be run concurrently, and offer brisk access to networked data and the Internet. Application design for desktops accommodates longer attention spans and more reliable data access. Application design for handheld devices sacrifices functionality in order to function in on-the-go environments where distractions are frequent, where batteries can run out, and where Internet connections are 'spotty'. Handheld devices are suited to in-the-field data entry, brief email checks, and management of personal information such as address books and calendars. Handheld use of the Internet is mostly suited to looking up current or near-future information such as entertainment and movie listings and driving directions. Evaluating the usability of designs for each device type requires different approaches for formal lab testing, but methods are similar for rapid prototype testing.

2

Handheld Devices

In this chapter, I introduce the three types of handheld devices: mobile telephone handsets, personal digital assistants (PDAs), and pagers. Next, I describe the hardware user interface elements and the most popular software user environments found on each device type. Finally, I elaborate on communication capabilities and detail the different wireless standards.

By 'handheld' I mean small enough to fit in one hand and without a necessary tether for either power or connectivity. 'Handheld' is critical in this definition, since it explicitly rules out laptops, printers and other peripheral devices, and standalone displays.

Connectivity is the ability to transfer data from one device to another, a network of devices, or the Internet. Connectivity is *not* required for a device to qualify as 'handheld'. However, in order for it to be described as 'wireless', connectivity *is* required. Devices therefore may be handheld but not wireless.

Device Types

The breakdown of handheld devices into three distinct categories – phones, pagers, and PDAs (see illustration 2.1) – may be considered controversial by some people. Each type of device in this breakdown has a primary use. As listed in Table 2.1 phones are used primarily for voice communication, pagers are primarily used for two-way email, and PDAs are primarily used for personal information storage and retrieval. As I mentioned in the previous

Illustration 2.1 Handheld device types: phone (left), pager (middle), and personal digital assistant (PDA; right)

Table 2.1 Differences between phones, pagers, and personal digital assistants (PDAs)

	Phone	Pager	PDA
Primary Use	Voice calls	Two-way email	Information storage and retrieval
Input Method	12-key keypad Stylus available on some communicators	Either no input or two-thumb keyboard	Varies: stylus, on-screen keypad, occasionally a two-thumb keyboard
Display Size	Typically very small, 12 characters by 4 lines is common; larger for communicators	Varies: from very tiny (one line of 16 characters) to 160 × 160 pixels or larger	160 × 160 pixels is common, but generally displays are fairly large
Form Factor	One-handed (tiny phones) or two-handed (communicators)	Smaller than a shirt pocket; about the size of a wallet	Typically fit in a shirt pocket
Communication Capabilities	Typically just voice, but via GSM, CDMA, GPRS, or other means; newer phones have cable connectors, infrared, and/or Bluetooth™	May only have GPRS capabilities, but some offer cable connectors, infrared, and/or Bluetooth™	Usually have a cable connector, but most have infrared and/or Bluetooth™
Expandability	Little or none; some phones allow snap-on hardware features, but this capability is rare	Little or none	Software expandability is extensive; additional applications can be added; many PDAs offer hardware add-ons as well

chapter, palmtops are not covered by this book. Palmtops are more similar to laptops than they are to any of the three device categories covered here, and so are best left out.

Form Factor

Form factor is the relationship between the dimensions of a device: its width, height, depth, weight, and configuration. Each of the devices considered in this book has many variations in form factor, but all are handheld, and all operate primarily without a power cord. Handheld device form factor differences can include display size, keyboard presence and orientation, whether a device has an integrated cover – and whether said integrated cover is merely protective or contains electronics or an additional display, etc. The form factor differences in handheld devices will only increase as the market tests new types of information appliances.

In the next three sections, I describe the three device types covered by this book, as well as the dominant user interface environments for each.

Phones

Mobile telephone handsets (phones) are used primarily for voice communications. Their secondary use is for SMSs (short message services). Their tertiary use is for WAP (wireless application protocol) or i-mode, two protocols for the mobile Internet. SMS, WAP, and i-mode are described in detail below.

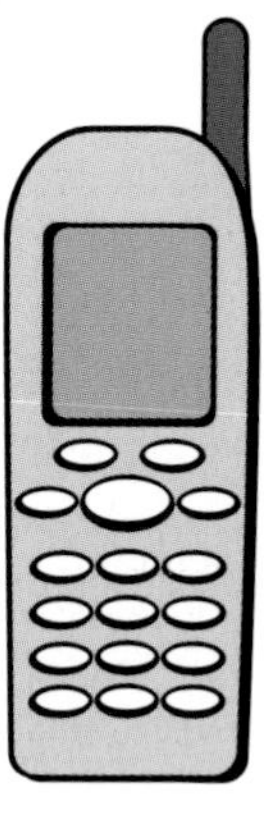

Illustration 2.2 Mobile phone

Mobile telephones come in several varieties, and have many different button sets. All handsets support at least 12 input keys, talk, end, and power (see Illustration 2.2). Most phones support 1 or 2 **softkeys**, whose functions change based on context. Softkey labels are always displayed immediately above them in the handset's display.

Handset displays range from very tiny – 16 characters by 4 lines – to fairly large – 160 × 160 pixels – as in Palm OS® PDA phones. Input on handsets is usually limited to voice and the keypad. New 'communicator' models feature larger displays and keyboards or stylus input.

Basic Phones vs. Communicators

Basic feature phones support voice, very simple contact management, and WAP. Communicators support the basic phone functions but also support a full suite of PDA features, such as custom applications, synchronization with a desktop computer, and Web browsing.

Phones come in a wide range of sizes and weights, from less than 3 ounces to more than 7 ounces. In size, the smallest phones compete with cigarette lighters, and the largest phones are affectionately referred to as 'bricks'. Phones may or may not have 'flips', a snap-open or snap-closed faceplate that accepts and terminates calls. Flips can cover the display and the keypad, or just part of either. Phones always fit into a typical back pocket or in a handy holster.

User Interface Environments

Most mobile telephone handsets have extremely customized user interface environments. Each phone manufacturer designs and constructs its own environment according to marketing requirements. The exceptions are Palm OS and Windows CE phones, whose user interfaces are extremely consistent with each operating environment's user interface standards. Perhaps the benefits of the Palm OS and Windows CE operating environments outweigh the benefits of uniqueness for these implementations.

WAP and i-mode are the most common ways to deploy applications for mobile handsets. The two are competing protocols, and one or the other is deployed on a given handset. For phones without WAP or i-mode, application development other than by the manufacturer depends on the underlying operating system – and the manufacturer's decision to allow applications to be installed on the handset.

WAP and i-mode

Wireless telephone service evolution has been described in 'generations', 1G, 2G, 2.5G, and 3G. 1G, now outdated, allowed analog voice communication only. 2G offered digital service, including caller identification and messaging, and Internet connectivity. 2.5G offers quicker data access via packet switching. 3G will offer packet switching and much higher speed connectivity, up to 2 Mbps (megabits per second). Today, most users worldwide are using 2G or 2.5G.

Mobile Internet connectivity is provided via one of two protocols: WAP or i-mode. WAP, the wireless application protocol, is the standard in Europe and North America, while i-mode is popular in Japan. Mobile Internet services enable users to send and receive email, get up-to-date weather, news, and transit information, and to purchase goods and services, such as movie tickets. Users interact with mobile Internet services through their telephone handsets.

WAP and i-mode are client–server protocols that are different from messaging. Messaging is a different client–server protocol, which I will address after discussing WAP and i-mode.

i-mode

i-mode is the world's success story for mobile data communication. It was developed as a packet-switched high-speed (9600 bps) system by Japan's dominant mobile telecommunications carrier, NTT DoCoMo. It is inexpensive, costing as little as ¥1 (less than 1 cent) to send an email message. Information about i-mode is available from www.nttdocomo.co.jp/english/p_s/imode. The technology and marketing of i-mode have made it a tremendous success, with more than 27 million subscribers as of 4 October 2001. This statistic comes from 'The Unofficial Independent imode FAQ', available from www.eurotechnology.com/imode/faq-gen.html.

Unlike voice calls, which are circuit-switched connections requiring dial up, i-mode's data connections are packet switched, meaning the data connection is 'always on'. There are certainly delays in the process, but they are brief compared with circuit-switched connections. i-mode charges according to the number of data packets sent to and from the telephone handset. Since the amount of data transferred is typically very small, incremental charges are tiny.

i-mode content is coded in cHTML, the compact hypertext markup language. cHTML is a subset of HTML but includes special tags that link content to telephone numbers that connect a voice call when clicked.

Messaging for i-mode users is straightforward. Internet email is integrated into the handset user interface and, as a result, email is the most popular i-mode application. i-mode email is limited to 250 double-byte characters. The Roman alphabet can be represented by single-byte characters, but Japanese requires double-byte characters, since the number of characters to represent is significantly greater. Additional characters are truncated, and attachments are not supported.

The popularity of i-mode is attributable to several factors. First, it is inexpensive to use. Second, access to it is easy, as phones that support i-mode have a dedicated button to access i-mode services. Third, i-mode is the primary means for most of its users to access email and the Internet, as desktop dial-up is costly in Japan. i-mode provides convenient access to email to people who might not otherwise have access to email at all.

WAP

WAP (wireless application protocol) is owned and maintained by the WAP Forum, a member-supported organization that has the world's largest telecommunications carriers as members. I use WAP as a generic description for services that utilize WML (wireless markup language) and HDML (handheld device markup language), WML's predecessor. WAP is presently a circuit-switched protocol that is charged by the time spent on a connection rather than by the amount of data transferred. As such, WAP is costly to use. WML and HDML are also more difficult to code than is HTML, making WAP an unappealing target for developers.

WAP was created for handheld devices, since the desktop Web was not suited to devices with small displays and extremely narrow bandwidth. However, WAP has failed to serve effectively the market for which it was created, because of marketing, reliability, and usability problems. WAP is expensive to use, since most carriers charge the same rates for minutes consumed during data calls as they charge for minutes used for voice calls. WAP is slow and tends to fail. While its data transmission speed is comparable to i-mode's (typically 9600 bps, according to Andrejs Jerkins and Ivaylo Todorov, in 'Critical Success Factors in the New Economy: How to Make WAP Worth IT; A Study of Wireless Internet'), its circuit-switched nature makes it very slow to connect. WAP's usability is very poor and is further compromised by the fact that most mobile telephone handsets hide its access behind cryptic menus or confusing icons. Most present WAP implementations do not support bitmap graphics or user interface 'widgets' such as menus, text entry fields, and text formatting, making the WAP experience alien to its users. Worse yet, WAP applications are extremely inconsistent with each other, requiring users to learn paradigms each time they try a new WAP user interface.

Multiple implementations

WAP browsers, unlike desktop Web browsers, are impossible to upgrade without replacing the hardware on which they reside. No protocol is in place for upgrading the software or firmware on mobile telephone handsets. If available, upgrades to firmware would require download and install software and protocols, which do not exist. Manufacturers have so far not developed upgrade software, perhaps because their revenue comes from consumers replacing their hardware. Alternatively, manufacturers could offer the upgrades for a fee, but it is unclear how much revenue such a strategy would generate. Therefore, it is extremely difficult to deliver the benefits of WAP browser improvements to consumers, who are understandably reluctant to discard their handsets each time a new version of WAP is released.

Since there are multiple versions of WAP, developers are forced to decide which versions to support, or to develop for the earliest version available to their target audience. Newer versions of WAP are backward compatible.

No standard button arrangements

WAP was designed to be adaptable to any small device. As such, no standards have emerged for handset button arrangements. Even worse, carriers have taken to writing WAP user interface guidelines on a device-by-device basis. So not only are there multiple devices to support, but each device may have different user interface guidelines from each carrier.

Fortunately, de facto standards have emerged, and some sets of user interface guidelines have been developed with a great deal of common sense. Sprint PCS publishes guidelines that are available from www.developer.sprintpcs.com. Sprint PCS's guidelines are especially useful, with specific examples for common WAP user interfaces.

Walled garden concept

While some WAP implementations allow direct dial of a WAP site via its corresponding telephone number, some implementations require access through a WAP connect feature on the telephone handset. Most WAP users are unaware of the direct dial feature that is sometimes available and rely on the handset's connect feature. The connect feature may be a dedicated 'M-Services' button (for mobile services), or a menu item accessed through a series of software menus embedded in the phone. More about M-Services can be found in the *M-Services Guidelines*, available from www.gsmworld.com.

Each telecommunications carrier controls the portal accessed via the 'Connect to WAP' handset feature. Companies who want their sites to be listed in a portal must negotiate with the carrier who owns it. Carriers often charge content providers a high fee for listing within the portal, even though the data calls generate revenue for the carriers. As a result, most carriers hide features that enable users to enter a website address directly.

Messaging for WAP users: short message service

Messaging in Europe and the USA is a messy affair. Unlike i-mode, which makes email very simple to use, the European and US systems only make SMS, the short message service, easy to access. SMS is limited to handset-to-handset communication, typically within a single carrier's network. The European community formed the GSM, the Global System for Mobile Communication, in 1987. GSM has all of the European carriers as members, and they agreed early on to link their SMS networks. Linking SMS networks has made SMS the 'killer' data application for all of Europe, since any European handset can send an SMS to any other European handset, regardless which carrier each uses.

SMS is wildly popular in Europe. SMS enables users to transfer short text-based messages between mobile telephone handsets at low cost. SMS messages require three fields: telephone number, subject, and message content. The popularity of SMSs is attributable to its marketing and its convenience. SMS is inexpensive to use, costing pennies (or pence) per message. Further, all European carriers are linked, so SMS works throughout Europe. SMS is also easy to use, despite the challenges of text entry on telephone keypads. A whole abbreviation language has emerged as a result, much like the language of license plates, where 'gr8' is short for 'great', and 'u' is short for 'you'.

Do not mix up SMS and WAP, however. SMS is only a message service, despite the amazing array of sophisticated applications that rely on it, such as vending machines and courier systems. It is also not 100% reliable, as messages can get lost or can take hours to be delivered. WAP is a rich client–server protocol for displaying and managing interactive content on handheld devices.

The user interfaces for SMS generation and display are unique to each handset, although they are remarkably consistent and easy to use. As full

)janogly Learning Resource Centre

Self-issue Receipt

04/02/08

02:50 pm

em: Handheld usability / Scott Weiss.
em ID: 1004625589

ue Date: 31/03/2008 23:59

The above item(s) may be recalled
before the displayed Due Date

keyboards on mobile telephone handsets proliferate, SMS usage should increase dramatically.

M-Services: the promise to come

The GSM released the *M-Services Guidelines* in May 2001 (www.gsmworld.com). The *Guidelines* specify a suite of features and services for new handsets. The key features of the *Guidelines* are listed in Table 2.2.

The 'M-Services key' is controversial, as it introduces yet another button to telephone handsets. However, having one-button access to the data features of a handset is compelling.

GPRS, the general packet radio service, is quickly becoming available in Europe. GPRS offers the promise of faster and less expensive WAP services through the always-on nature of packet-switched data networks.

The graphical Web browser offers usability enhancements, such as standardized user interface widgets. It also will support bitmap graphics and audio. The data demands of the graphics will be offset by the always-on nature of GPRS.

File system features will enable users to download content to their handsets directly through WAP. Ring tones, background screen images, and games will be the first content.

MMS is the multimedia message service, a new standard developed by the GSM. It allows for longer handset-to-handset messages, as well as additional content, such as bitmaps, audio, and video. It is a shame that POP3 email (Post Office Protocol version 3, the standard for Internet-based email) was not required, instead, as POP3 email is the global standard. In the *M-Services Guidelines* POP3 email is specified as 'optional'.

Table 2.2
Key features of the *M-Services Guidelines* (available at www.gsmworld.com)

Feature	Benefit
M-Services key	Ease of locating WAP services
GPRS (general packet radio service)	Always-on data connectivity
Graphical Web browser	Familiar user interface widgets and bitmap graphics
File system features	Ability to download and manage ring tones and 'wallpapers' from the network
Enhanced messaging	MMS (multimedia messaging service) and (optionally) POP3 email

WAP will continue to catch on, especially as the features of the *M-Services Guidelines* are implemented in mobile handsets and networks.

Symbian OS

All mobile telephone handsets require an underlying operating system (OS), but until recently, each manufacturer built or licensed a custom OS. Symbian OS is making inroads into the handset market, although it faces stiff competition from Palm and Microsoft. OS offerings from Palm and Microsoft will be discussed in the section on PDAs, below.

Symbian presently markets Symbian OS as an operating system designed for the specific demands of mobile telephone handsets. Symbian released the world's first mass-produced handheld computer, the Psion 1, in 1984. Symbian OS now runs on mobile telephones and Psion palmtops. On telephone handsets made by Ericsson and Nokia (see Illustrations 2.3 and 2.4), the Symbian

Illustration 2.3 Ericsson R380e phone. Reproduced with permission from Ericsson

Illustration 2.4 Nokia 9290 communicator phone. Reproduced with permission from Nokia

OS look-and-feel is presently hidden under licensees' own look-and-feels.

Both of the phones pictured use Symbian OS. They offer rich user interfaces that are designed for mobile professionals, but each device's look-and-feel is completely different. Both phones support WAP, and Ericsson and Nokia both offer user interface guidelines for WAP design. Nokia also offers user interface guidelines for developing Symbian OS applications for the 9290 Communicator, but Ericsson does not reveal Symbian OS to developers, instead requiring them to use WAP as the only development platform for the device.

There are strong advantages to development in Symbian OS over WAP. While WAP uses a variant of HTML, Symbian OS features C++ and Java development. C++ is a sophisticated object-oriented programming language that offers developers many more opportunities than WAP. Java is an open software platform from Sun Microsystems Inc., which promises developers the ability to develop applications once and deploy them on a variety of platforms.

The key disadvantage to developing in Symbian OS is that the number of handsets able to download and run a given application is far fewer than the number of handsets able to run WAP applications. Nokia markets the 9290 Communicator and 7650, two radically different handsets. The 9290, as pictured, is a communicator phone with PDA capabilities and a QWERTY keyboard. The 7650 is a more traditional looking handset with a 12-digit keypad, but features a digital camera. It is not clear that an application developed for the 9290 will look and feel appropriate on the 7650, or if the application will even run without modification. Ericsson uses Symbian OS for the R380e, but developers cannot take advantage of it, as no developer access to the OS is available.

Pagers

Pagers are handheld devices used primarily for two-way email communication (see Illustration 2.5), but also feature some PDA capabilities, such as contact management and a calendar. I distinguish pagers from **beepers**, which are one-way communication devices with single-line displays; beepers are not addressed by this book. Some pagers also offer Web browsing. Pagers have QWERTY keypads suitable for 'two-thumb' typing. QWERTY keyboards are the peripherals attached to desktop computers that are suitable for touch typing. QWERTY keypads are a tiny variant suitable for handheld devices. The distinguishing factor between a pager and a PDA is the touch screen, which makes a fundamental difference in how applications are designed.

Illustration 2.5 Pager

Illustration 2.6 RIM 850™. Reproduced with permission from Research in Motion Ltd.

The two most popular pagers today are the RIM Wireless Handhelds™ and the Motorola Timeport® P935 (see Illustrations 2.6 and 2.7). RIM Wireless Handhelds™ have a proprietary operating system that I will refer to as RIM OS. Motorola's pagers run Wisdom™ OS, which is also available on the Accompli 009 mobile telephone.

Pagers are increasingly offering additional features, such as Web browsing and voice capabilities. Web browsing on a pager is somewhat frustrating, since in order to interact with user interface elements, one has to move a highlight focus with a directional keypad. However, the same frustration occurs with mobile telephones that support WAP.

Some pagers support WAP, while others have HTML Web browsers that use 'web clipping'. Web clipping is the act of stripping content from web pages that is unsuitable to a handheld device. In some cases, bitmap graphics, color information, and text formatting are 'clipped'.

Both RIM Wireless Handhelds™ and the Motorola Timeport® support application downloads. Developing for each platform is independent and must be done in the device's corresponding operating system. The RIM OS and Wisdom™ OS user interfaces are described in detail in the chapter on *Information Architecture: Practice* (Chapter 4).

Illustration 2.7 Motorola Timeport® P935. Reproduced with permission from Motorola Inc.

PDAs

PDAs are feature-rich stand-alone devices that have address books and calendar functions (see Illustration 2.8). All PDAs have touch screens and use a stylus for input, although some PDAs are coming on the market with QWERTY keypads and styli. Most PDAs also support email, to-do lists, note taking, and desktop synchronization. Most PDAs allow developers to write native applications for them in their respective operating systems. Newer PDAs feature Internet connectivity via internal or add-on hardware and software.

PDAs are typically larger and heavier than phones. PDAs range from just over 4 ounces to over 7 ounces. They come in vertical or horizontal layouts, as clamshell or always-open models. The largest PDAs are slightly smaller than palmtops, while the smallest are only slightly larger than the largest mobile phones. Palmtops are *not* PDAs, although they closely resemble them. Palmtops have larger displays and are best used on a table rather than in one's hands.

Illustration 2.8 Personal digital assistant (PDA)

User Interface Environments

The two most popular user interface environments for PDAs are Palm OS and Windows CE. There is stiff competition between these two platforms, despite the radical difference in their development philosophy. Windows CE takes the desktop user interface and shrinks it to fit in a handheld device, while the Palm OS environment was designed as a PDA user interface from the outset, taking only the features absolutely necessary to include in a handheld device and jettisoning the rest.

Palm OS®

The Palm OS platform revolutionized the handheld device industry in 1996 with the release of the Palm™ Pilot handheld. The Palm team looked at the available PDAs, mainly Apple's Newton, and stripped away everything from the design that they could. They reduced power consumption and hardware, making it possible to price the device much lower than the Newton. They made the inspired decision to position the device as an 'information appliance', rather than a tiny computer. Of course the Palm devices really *were* tiny computers, albeit less functional computers than their desktop brethren.

Modern-day Palm OS® devices have up to 16 Megabytes of RAM, color displays, and some even feature QWERTY keypads. However, the Palm OS main operating environment still looks and feels like the first version. It is simple to use, but it is functionally limited: it lacks a central file system, which requires users to look for data application by application, rather than from one dedicated location. The lack of a central file system also makes housekeeping a particular chore on Palm OS® devices, since there is no one place to look to delete unwanted files. Software vendor DataViz® has extended Palm OS with the Documents To Go product, providing some file system capabilities, but even that functionality is limited to just the files managed by Documents To Go. The Documents To Go product manages spreadsheets, word processing documents, and slideshow presentations within a central file listing. Sheet To Go is shown in Illustration 2.9.

Most Palm OS® devices have a 160 × 160 pixel display. The size of the pixels varies from device to device, ranging from 0.28 mm/pixel to 0.32 mm/pixel. Palm OS software can be found in PDAs and communicators.

Palm OS devices rely on a touch screen, which is used with a stylus. The touch screen on most Palm OS devices has a dedicated area for character recognition. The Handspring Treo™ 180 lacks this dedicated area, but third-party software enables character recognition anywhere on the display.

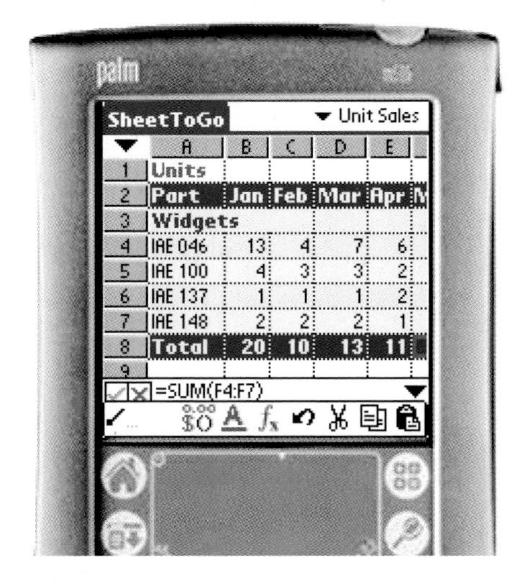

Illustration 2.9 DataViz® Sheet To Go for Palm OS. Reproduced with permission from DataViz Inc.

Microsoft Windows CE

Microsoft has written one operating system for all 'consumer electronics' devices, called Windows CE. This product is deployed in three flavors: Pocket PC, Handheld PC, and Smart Phone. Windows CE is a 32-bit multitasking operating system scaled for handheld devices. However, it operates like a miniaturized version of Windows, making it feel cumbersome to use on a small display without a full keyboard. Windows CE devices have shorter battery life and tend to be more expensive than Palm OS devices, but they also tend to be more powerful. For example, Windows CE for Pocket PC supports 240 × 320 pixel displays, larger than those available for Palm OS devices.

Windows CE supports a central file system, enabling users to do housekeeping chores easily and enabling application developers to rely on this user interface to manage their documents effectively. Windows CE applications feature a prominent menu bar, positioned at the bottom of the display (see Illustration 2.10).

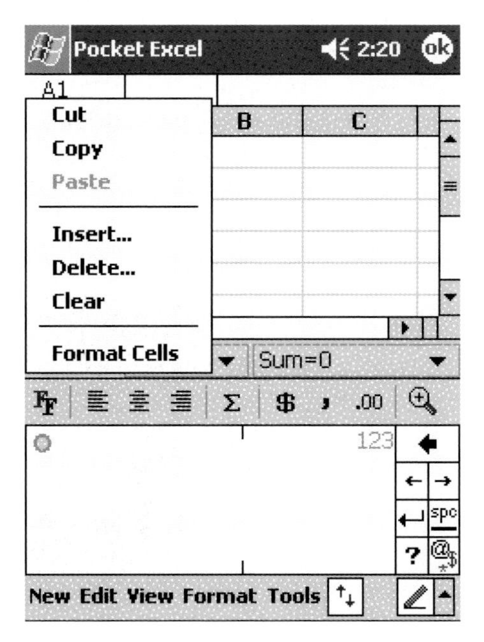

Illustration 2.10 Pocket Excel. Reproduced with permission from Microsoft

Communicators

While there have been increasing numbers of combination devices hitting the market, handheld devices still fall into three categories: telephone handsets, PDAs, and pagers. If a device combines the features of a phone and a PDA, it must follow the user interface design principles of both devices, which makes it much harder to design, but not harder to design *for:* design for the PDA, but integrate that design with phone capabilities.

There are different design considerations for each of these devices, but there are, of course, areas in which design considerations converge. Each of these classes of device has a small display and problematic data input. The device pictured in Illustration 2.11, the Handspring Treo™, attempts to beat all odds by providing a QWERTY keypad, 'large' display, voice communication, and Palm OS capabilities. The device overcomes the Palm OS platform's reliance on Graffiti® text input by providing a keyboard, although the Treo™ is also available with Graffiti input instead.

Combination devices are getting stranger and stranger (see Illustration 2.12). There are pagers with headset jacks, PDAs that double as phones, and phones with PDA capabilities. Combination devices rarely offer the 'best of both worlds', and all involve tradeoffs. Some PDA phones are heavy and bulky, and the displays may be slightly smaller than in the standalone PDA model. Some pager phones are big and heavy – not pager-like qualities.

Illustration 2.11 Handspring Treo™ communicator, which runs Palm OS. Reproduced with permission from Handspring

Illustration 2.12 Nokia 5510 communicator with MP3 player. Reproduced with permission from Nokia

However, the allure of carrying one fewer – or only one device – is extremely compelling.

Hardware User Interface Elements

All handheld electronic devices have displays, labeling, and buttons. Some devices have silk-screened buttons and softkeys. This section describes each of these user interface types and offers guidelines for their use.

Labeled Buttons

All handheld devices support labeled buttons of some type (see Illustration 2.13). Buttons typically contain silk-screened labels, although some button labels are positioned adjacent to the button itself.

Recommendations

- Buttons should have an obvious 'click' feel. One should know when the button has been pressed.
- The button label should be easy to read and understand. It should be screened onto the button using a high contrast color or luminance. The label should be as large as possible, since many users will have a hard time reading it.

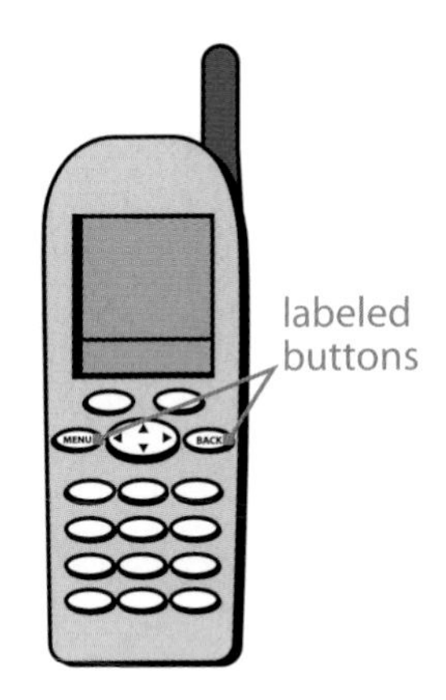

Illustration 2.13 Labeled buttons

- Icons should be obvious and cross-cultural. It is rare that keypads are custom-made for different regions.
- Button layouts should follow standards. While telephone keypad standards are hard to find and weak, the 12-button keypad is ubiquitous. Make sure the orientation is correct, with '1', '2', and '3' left-to-right at top, and '*', '0', and '#' left-to-right at bottom.
- For devices that include styli, place an indentation in the button where the stylus will 'hit' so that users who so wish can press the button with their stylus. Palm devices have such an indentation in the power button, but not in the other buttons.
- Button presses should produce obvious audio and/or visual feedback.
- While it is uncommon for 'Shift' buttons to work as they do on typewriters, some users will expect this behavior. Consider supporting it, in order to meet user expectations. When a typewriter 'Shift' key is held down, the next key is 'shifted' so that the alternate character is typed. Most handheld 'Shift' keys work as a stepped function and do not support simultaneous pressing.
- Button presses should initiate on the down stroke, not on the up stroke.

Softkeys

Softkey labels typically appear immediately above them on a changeable display (see Illustration 2.14). That display may be separate from the main

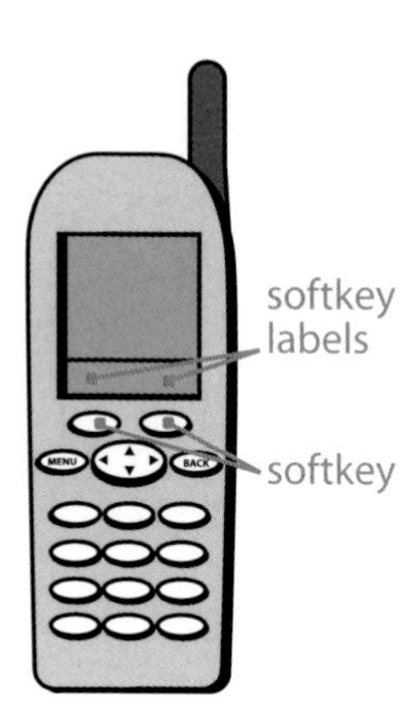

Illustration 2.14 Softkeys and softkey labels

device display. Softkey behavior should be modeled in exactly the same fashion as for labeled buttons. Devices that have touch screens usually do not have softkeys, since the display itself works like many softkeys. What would be softkeys end up being on-screen buttons. However, hardware keys are easier to press than on-screen buttons.

Recommendations

- Softkey labels should never scroll. They are labels, not information displays.
- The standard is for phones to support two softkeys. The GSM *M-Services Guidelines* require only a single softkey, since some handset manufacturers feel that only one is appropriate. The 'number of softkeys' issue is similar to the 'number of buttons on a mouse' issue. Microsoft Windows requires two, one for action and the other for a popup context menu. Apple has only one button on its mice and uses a keyboard modifier to produce a context menu, when available. However, there is no opportunity for a keyboard modifier on a telephone handset, so designing applications where the second softkey is always optional poses a significant challenge.
- The leftmost softkey should be the primary, or 'action', softkey. Some manufacturers have implemented the rightmost softkey as the action key, which is a shame, since users of those nonstandard phones will have to adapt to a nonstandard user interface.

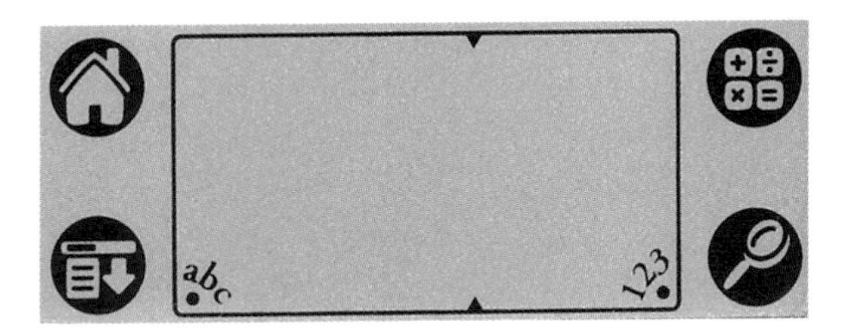

Illustration 2.15 Palm OS® silk-screened buttons. Reproduced with permission from Palm Inc.

Silk-screened Buttons

Silk-screened buttons are graphics or text that are printed on or under the glass of a touch screen display (see Illustration 2.15). These buttons are much harder to press than hardware keys, but their forced consistency (i.e. they are always visible and always look exactly the same) makes them more usable than on-screen graphical buttons, which have different placements, sizes, and appearances.

Navigation Controls

Directional Keypads

Most devices have two or four directional arrow buttons, as shown in Illustration 2.16. The controls move the highlight and/or they scroll content. Some implementations use discrete buttons, while others use a floating pad that can be pushed in only one direction at a time.

Illustration 2.16 Two-way and four-way directional keypads

Roller Wheels

Roller wheels are offered on the RIM Wireless Handhelds™, Sony Clie, and some Nokia mobile telephone handsets. Roller wheels are used to scroll through menu options and frequently can be pushed to activate a selection button (see Illustration 2.17). Roller wheels are called different things by different manufacturers, including Nokia's 'naviroller', Sony's 'jog-dial', and RIM's 'trackwheel'.

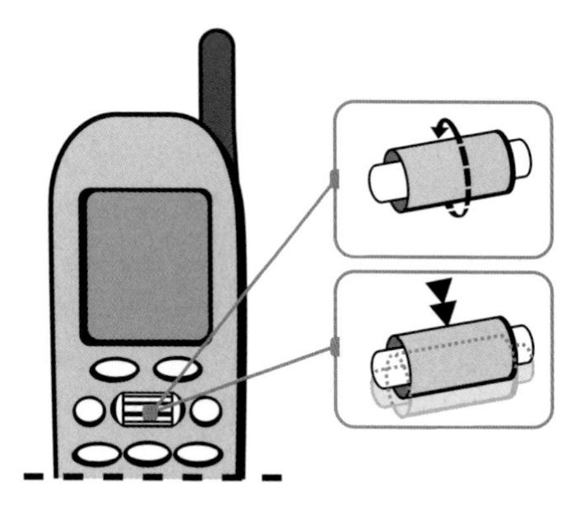

Illustration 2.17 Roller wheel

Some roller wheels have additional manipulation options that enable more complicated navigation. However, these additional manipulation options seem to hurt usability more than help it, as the rolling and clicking mechanisms offered by most are fairly complex already.

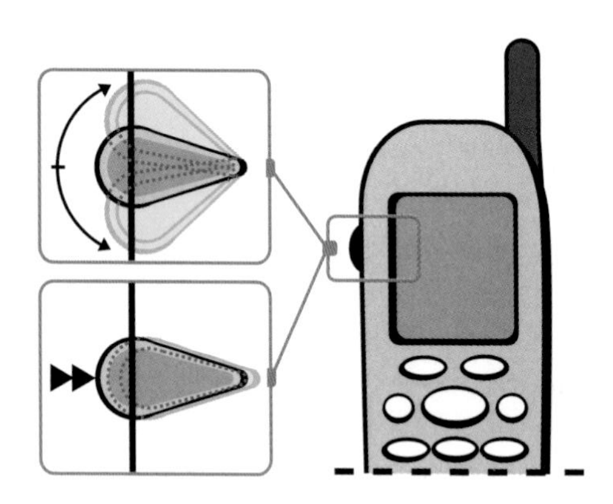

Illustration 2.18 Rocker control

Rocker Controls

Rocker controls are directional controls that are usually placed at the upper left-hand side of a device (Illustration 2.18). These controls can be pushed up and down, and most rocker controls can be clicked 'in' as an additional manipulation option. Rocker controls are used for scrolling, selection, and action. With some rocker controls, the farther in a direction the control is pushed, the faster the scrolling goes. With others, the longer the rocker wheel is pressed, the faster the scrolling action.

Rocker controls look and behave much like roller wheels, but rocker controls do not 'spin'; they stop; though the software keeps incrementing, as when scrolling within a list box.

Touch Screens

Touch screens enable point-and-click with a finger or stylus. Owing to the thickness of the plastic of the display, parallax is a common problem: where one thinks one is clicking may not be where the click appears on the display. Touch screens are not a substitute for mouse input, as they are not as accurate as a mouse. Mice can track up to 400 points per inch, while touch screens are only as accurate as the tip of the stylus – even less so if parallax problems occur.

Many PDA–phone combination devices are becoming available that lack a 12-digit keypad and instead rely on a touch screen keypad. These devices sacrifice tactile feedback and require their users to look at the display while dialing.

Data Input

All of the data input mechanisms for handheld devices are suboptimal. One manufacturer, Rolodex, made a device that did not even *support* input. The ill-fated REX® Micro PDA enabled its users to display their calendars and address books but not to make updates with the handheld. In order to update, users had to enter information on a desktop computer and then synchronize the device. I had one and carried around white cards on which I wrote my schedule changes and address book additions. However, the device did not catch on and has since been discontinued.

None of the handheld device data input options are ideal; each method involves tradeoffs. This section details the available technologies and describes their strengths and weaknesses.

Keypad Input Options

There are many different methods for entering text data into handheld devices. All of these methods require learning, and some of them have a greater learning curve than others. As an information architecture designer, you may not have control over the data input mechanism, but you should have a working knowledge of the different methods. Having this knowledge will enable you to design user interfaces that are optimized for the available text input options.

Triple Tap and T9®

Triple tap is the act of pressing a number key up to four times in order to input a single text character. For example, the '2' key is pressed four times to input the letter 'C' (2, A, B, C). This input method is the default for all telephones when text input is required. There is no copyright or patent on this method, and it is deployed on every telephone handset. However, it is difficult to use and is not intuitive.

AOL Mobile's T9® is a predictive text input technology that uses a numeric keypad to enter text, one keypress per character. However, the display shows what the predictive interpreter foresees as the complete word after each keypress. In our usability research, users were initially confounded by the dichotomy between saving keystrokes and displaying confusing results. Ultimately, respondents were no quicker with T9® than with triple tap, although some respondents were fairly excited about the technology. The problem with T9® goes beyond the learning curve; what shows up on screen is confusing, pure and simple. In Table 2.3 the sequence required to produce the word 'meeting' on the T9® is compared with that with triple-top input.

Stylus Input Methods

The stylus, a pointing device that looks like a pen, affords several different input methods. All systems that support styli have receptacles where the stylus can be stored.

Table 2.3 The sequence required to produce the word 'meeting'

Triple-tap input		T9® input	
Key pressed	Display	Key pressed	Display
6	m	6	o
3	md	3	of
3	me	3	off
pause	me	8	meet
3	med	4	meeti
3	mee	6	meetin
8	meet	4	meeting
4	meetg		
4	meeth		
4	meeti		
6	meetim		
6	meetin		
4	meeting		

Electronic ink

Electronic ink echoes the stylus strokes on the display, a direct machine duplication of what an ink pen would produce on paper. Electronic ink is useful for simple drawings and text notes.

Handwriting recognition

Handwriting recognition can recognize individual characters, either single-stroke or multistroke. For example, 'C' is always drawn as a single stroke, but 'A' has two strokes – the inverted 'v' and the cross bar. More sophisticated systems recognize whole words. Some systems require users to enter characters in a dedicated area on the display, while others allow character entry anywhere on the display.

Graffiti® Most PDAs support stylus input. A stylus is used to write or 'tap' a touch screen display. Most Palm OS handhelds have a dedicated area of the screen for handwriting recognition, but other systems allow writing anywhere on the display.

Graffiti software is a proprietary handwriting recognition technology owned by Palm. It enables users to 'hand write' characters on the display with a single stroke, as shown in Illustration 2.19. All Palm OS devices support

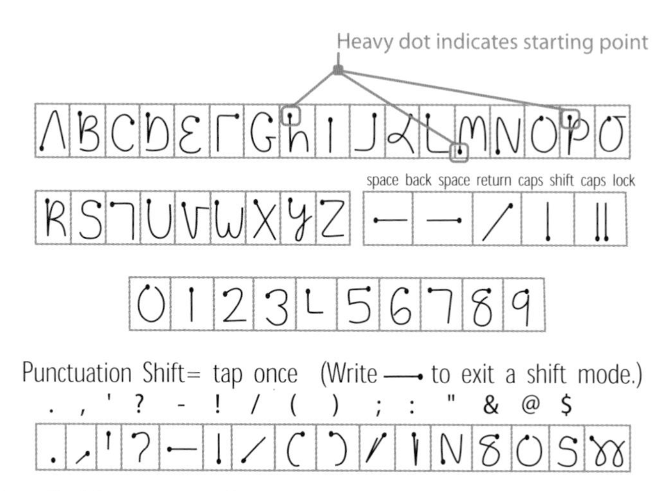

Illustration 2.19 Palm OS® Graffiti® guide. Reproduced with permission from Palm Inc.

Graffiti input, including the Kyocera 6035 telephone handset. While some users become quite deft with Graffiti input, most become merely competent, hence the demand for portable keyboards.

CIC's Jot® CIC (Communication Intelligence Corporation) produces Jot®, a natural handwriting recognition software product. Jot® recognizes multiple forms for each character, including both single-stroke and multi-stroke styles

Illustration 2.20 CIC Jot® stroke options for the character 'a'. Reproduced with permission from CIC

(see Illustration 2.20). Jot® is included on Windows CE devices and is available for purchase on Palm OS devices. The key disadvantage to Jot® is that it requires more computing power than Graffiti software, and so using it reduces battery life and available RAM.

On-screen keyboards

Some users will not be able to write with Graffiti, T9®, or even triple-tap input

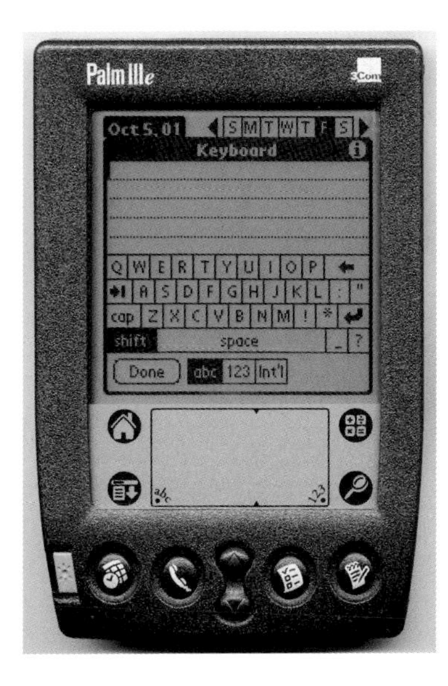

Illustration 2.21 Palm OS® platform's on-screen keyboard. Reproduced with permission from Palm Inc.

methods. For that reason, devices with large displays offer on-screen keyboards, as shown in Illustration 2.21. The on-screen variants do not support touch typing, as one might expect. They instead require users to tap each character with the stylus – a slow and deliberate process.

Digit Wireless's Fastap™

Fastap™ is a full keyboard deployed on a telephone handset. This technology is quite promising, as it may make it far easier for users to enter text with minimal difficulty – without requiring a full-size keyboard. Digit Wireless even has a QWERTY version of their design, which is smaller than a credit card.

Illustration 2.22 shows how the Fastap™ keypad works. It relies on a 'hills and valleys' model; alphabetic characters are the hills, and numeric characters are the valleys. Clicking a hill produces that key, but the valleys are little more than a rubber 'plate' linking the four surrounding hills. Press a valley and you end up pressing the four surrounding hills. As you might expect, pressing a valley is a bit harder than pressing a hill, but since the valleys are larger in size, the effect is pleasing to the eye and makes for a surprisingly easy-to-use entry method.

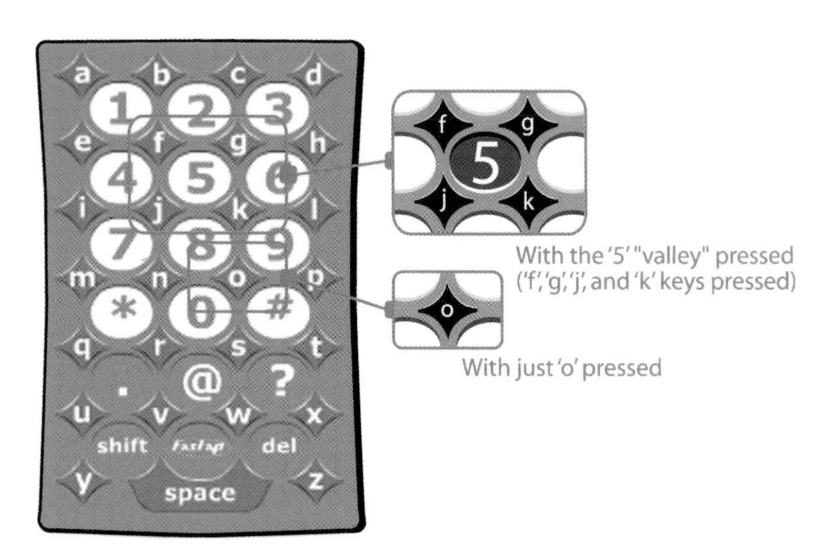

Illustration 2.22 Digit Wireless's Fastap™. Reproduced with permission from Digit Wireless

QWERTY Keypads and Keyboards

QWERTY is named for the left-to-right arrangement of keys on an English keyboard: Q-W-E-R-T-Y. Keypads have tiny buttons, whereas keyboards have large buttons that enable touch typing. QWERTY keypads are included on pagers, and QWERTY keyboards are available as accessories for some PDAs, as shown in Illustration 2.23.

QWERTY keyboards, no matter how small, are the easiest input mechanism for handheld devices. They can be available as folding keyboards, as shown in Illustration 2.24.

Despite their popularity, you cannot guarantee the availability of a QWERTY keyboard – not everyone will purchase one. Even with the best character input technology, however, do not expect your users to write their next novel on their handheld device.

Displays

All handheld devices have one thing in common: small screen size. Displays vary with regard to dimensions, ability to display bitmaps, color resolution, and touch or stylus sensitivity.

Illustration 2.23 The RIM BlackBerry™ (left) and Motorola's Timeport® P935 (right) have QWERTY keypads. Image of BlackBerry™ reproduced with permission from Research in Motion Ltd. Image of Timeport® P935 reproduced with permission from Motorola Inc.

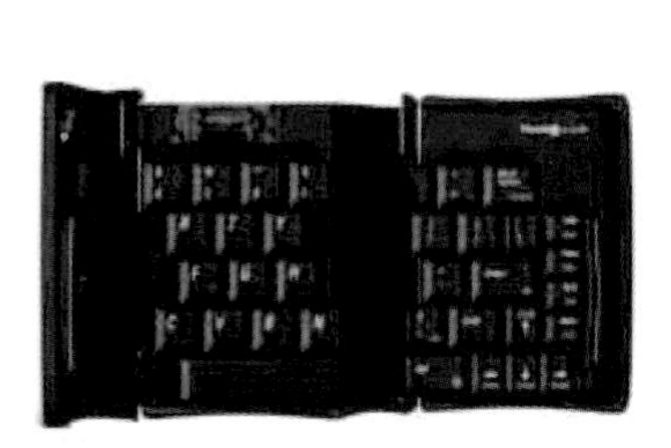

Illustration 2.24 Think Outside offers a QWERTY keyboard that worked with PDAs and cell phones. Reproduced with permission from Think Outside

Size and Resolution

There are two types of display sizes: 'small' and 'PDA-style'. Small displays tend to have fewer than 6 lines, with up to 20 characters per line. PDA-style displays can be as large as 160 × 160 pixels (Illustration 2.25), which can result in up to 20 lines of 20 characters per line, but not necessarily – characters can be as small as 5 × 7 pixels or as large as 20 × 20. Although PDA-style phones may support the same pixel resolution as plain PDAs, their displays are physically smaller, resulting in smaller pixels.

When designing phone applications, you simply cannot predict the resolutions of the phones on which your application will appear. My recommendation is to design for 4 × 12 characters, including softkey labels. Most displays are capable of displaying more characters, but if your application is to work on all telephone handsets, you must design for the least common denominator.

Researching the resolution of the devices you plan to support is your first challenge; some examples are shown in Illustration 2.26. Sprint PCS recommends supporting 12 characters by 4 lines. However, there are phones with 3 lines, so even Sprint PCS' conservative estimate may not be adequate. On the flip side, Motorola's Accompli™ 008/6288 supports 240 × 320 pixels. Pixel-to-character conversions are tricky at best, since most devices support different type sizes.

Most devices support only black and white, but color displays are emerging quickly. As cost decreases and battery technology improves, more color devices will appear. Most phone and pager displays are text-only, but that fact is

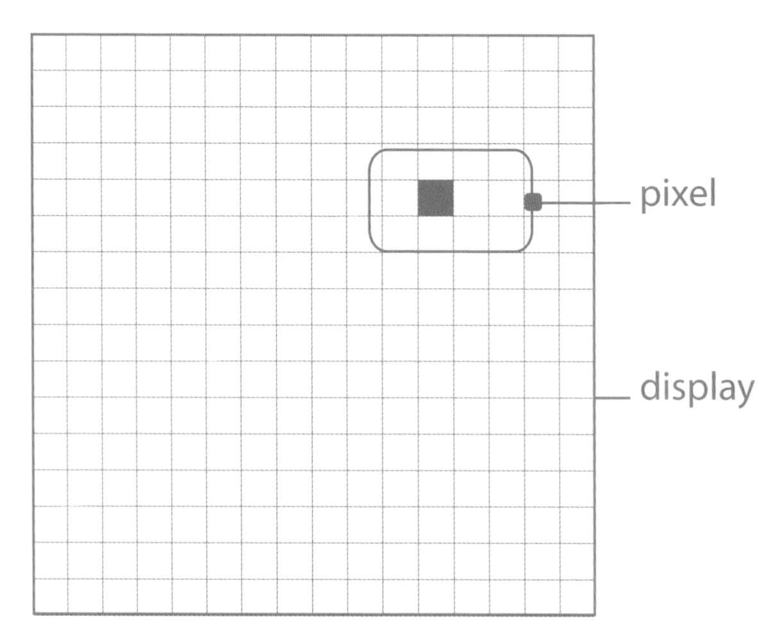

Illustration 2.25 Display size in relation to pixel size

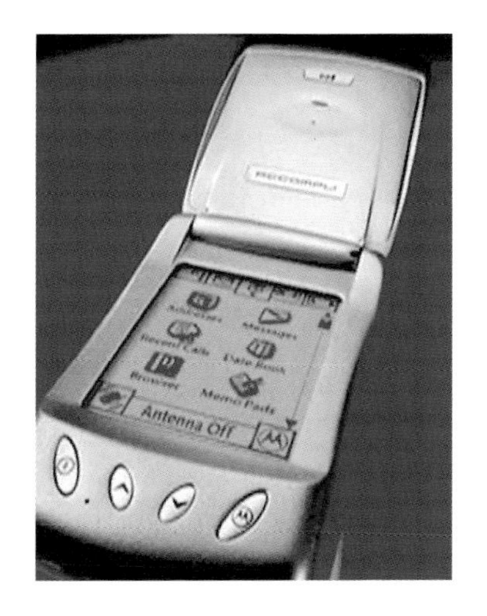

Illustration 2.26 Display sizes in use: (left) Nokia 8890, a 'small' display; (center) Motorola's Accompli® 008, and (right) a 'PDA-style' display; Wireless Handhelds™ (right). Image of Nokia 8890 reproduced with permission from Nokia. Image of Accompli™ 008 reproduced with permission of Motorola Inc. Image of RIM Wireless Handhelds™ reproduced with permission of Research in Motion Ltd.

changing quickly. Early WAP browsers do not support bitmapped graphics, though, so keep this fact in mind when designing WAP sites.

Design for Small Screens

Small screens are difficult to use and even more difficult to design for. The fundamental requirement is that content be contiguous, without blank lines. The most important content, naturally, should appear at the top of the page. Never use blank lines on a small display, since the user may interpret the blank line as the end of the page. Use dashes ('-') to create separations in content.

Expansion

Most devices in the handheld category offer cradle connections to desktop computers for synchronization and file downloads. Cradles are not for expansion, however. Expansion is when a hardware add-on is affixed to the device by way of a connector port. Cameras, scanners, memory add-ons, modems, phones, and other devices can be added via expansion slots.

Feedback when installing an expansion device should be three-fold:

- Tactile: there should be a satisfying 'click' when attaching an add-on device.
- Aural: the device should chime with an 'optimistic' tone when the add-on is recognized.
- Visual: the device should display a screen message indicating the add-on has been recognized.

Configuration is often necessary when add-ons are attached. Configuration screens should be immediate, but allow for postponement. Either way, a brief introduction to the add-on should be offered, stating:

- what the add-on can do for the user
- how the add-on can be used
- where to get additional information about the add-on

Connectivity

Most connectivity today relies on cable and Ethernet jacks. Ethernet is a standard that was invented by Xerox and is now used by every computer manufacturer. Connectivity between desktop computers is called 'peer-to-peer' networking. When intermediate devices, called **hubs** or **routers**, are added to purify and strengthen the signals (so that they can travel longer distances and connect larger numbers of computers) the system is called a LAN, or local area network. LANs are usually deployed over multiple floors of a building or between adjacent buildings. WLAN describes wireless LAN. Wireless connectivity is becoming increasingly popular, yet security considerations and ease-of-use problems with authentication have been contributing to challenging this technology's acceptance. Beyond LAN, there is WAN, wide area networking. WANs are essentially multiple interconnected LANs, covering large distances, such as miles or hundreds of miles. WANs may rely on long-distance cable or even on satellite connections for data transfer.

Authentication, the process of identifying and validating a user, is more complex with wireless than with wired connectivity. Let's say you walk into a meeting with your handheld. You want to beam an appointment to your colleague, and your handheld uses 802.11b, a radio frequency standard that does not rely on line of sight. How does your handheld know the identity of the handheld to which you want to beam, and how does the receiving handheld know to listen for your signal? Even worse, how can you keep the other meeting attendees' handhelds from capturing your data transfer? These issues have not been effectively worked out yet for consumer devices.

Security and competing networks are two other challenges. Competing networks within the same physical location pose similar problems to those encountered by millions of portable home telephone users every day: reduced throughput or terminated connections resulting from interference from closely located, separate networks.

TCP/IP

A US Department of Defense research project developed TCP/IP, Transfer Control Protocol/Internet Protocol. It is the world standard for computer networking. A 'network of networks', TCP/IP relies on packet switching between computers and the follow-up verification of the data. It is an extremely robust system that automatically recovers from most failures. The Internet relies on TCP/IP.

Computer addressing in TCP/IP can be dedicated or dynamic [dynamic host configuration protocol (DHCP)]. Either way, an IP address is a set of four numbers between 0 and 255, separated by periods. All devices on the Internet must have an assigned address or share one by use of a router.

Internet

Internet connectivity is basically the ability to access the Web. However, the standard that the device supports is also relevant. Internet standards include HTTP (hypertext transfer protocol), which is the desktop Web, and WAP (wireless application protocol), the most common standard for mobile handsets. FTP (file transfer protocol), XML (extensible markup language), SMTP (simple mail transfer protocol), and other protocols exist but are outside the purview of this book.

Intra-device

Intra-device connectivity addresses transferring data between two handheld devices, between handheld devices and desktop computers, or between handheld devices and peripherals, such as printers. Most of us have experience with beaming or infrared data transfer pioneered by the Palm OS platform, but intra-device wireless communication is becoming more available through the increase in popularity of IEEE 802.11b, from the US Institute of Electrical and Electronic Engineers.

Types of Wireless Connectivity

Emerging wireless connectivity standards are intended to work between many types of devices, both tethered and untethered. Connectivity is provided either by radio frequency or infrared and offers different distances between repeaters, depending on the nature of the standard. All wireless local area connectivity requires a **base station** that plugs into an electrical outlet and, if it offers Internet connectivity, it must be wired into a service provider, as well.

Infrared

Infrared and the Infrared Data Association (IrDA) standard rely on a narrow portion of the light spectrum and require line-of-sight in order to work: infrared does not go through walls. It is present in nearly all Palm OS devices. The Palm OS platform relies on **beaming** to transmit information between devices. Beaming requires close proximity and has a fairly uncomplicated authentication process, requiring the beam recipient merely to accept the data transfer.

While infrared transfer has been incorporated into many printers, laptops, and digital cameras, its line-of-site limitation is significant. Radio frequency technologies, such as Bluetooth™ and IEEE 802.11, are likely to eclipse it in popularity.

Bluetooth™

Bluetooth™ is a radio frequency connectivity standard for handheld devices, not for local area networks (LANs). The standard allows for connectivity within 30 feet (approximately 10 meters) and has no specification for user interface. Bluetooth™ was designed for low-cost and low-power consumption, making it ideal for mobile telephone handsets and PDAs. The Bluetooth™ wireless networking standard is endorsed by a consortium of companies: 3Com, Agere, Ericsson, IBM, Intel, Microsoft, Motorola, Nokia, and Toshiba are members of the Bluetooth™ consortium. Information about Bluetooth™ is available from www.bluetooth.org and www.bluetooth.com.

IEEE 802.11a, b and g

Otherwise known as Wi-Fi (Wireless Fidelity) or Apple's AirPort, this wireless standard is rapidly becoming the most popular on Windows PCs and Macintoshes. It is directly modeled after Ethernet and so is very compatible with existing wired networks of desktop computers, printers, and other peripherals. Although 802.11 is designed for LANs, it is quickly being incorporated into PDA designs. IEEE 802.11 is a high-speed standard, supporting up to 54 Mbps (megabits per second) – much faster than Bluetooth™. 802.11 devices can communicate much farther apart than Bluetooth™ devices – up to hundreds of feet. IEEE 802.11 Direct Sequence Spread Spectrum (DSSS) connectivity is becoming available in coffee houses and even community-based bandwidth sharing programs. Free wireless Internet zones are being established all over the world through grass-roots efforts to share excess bandwidth.

The 802.11g standard was approved on 16 November 2001 by the Institute of Electrical and Electronic Engineers (IEEE). This new standard is backward-compatible with 802.11b, the previous standard. 802.11g supports wireless data transfers of up to 54 megabits per second, compared with 11 megabits per second with 802.11b. The earlier standard, 802.11a, operates in the 5 GHz range, and so is incompatible with 802.11b.

Both Bluetooth™ and 802.11g operate in the 2.4 GHz communication band, which is also used by newer portable home telephones. Devices supporting both Bluetooth™ and 802.11b are becoming available, and I suspect that 802.11g will be widely adopted.

Mobile telephony and data standards: cellular, GSM, GPRS, and others

Mobile telephony standards enable handsets to operate above and below ground, in vehicles, and even in the air (despite not being allowed in planes) – with varying levels of reliability. Mobile telephony relies on 'cells', or land-based antennae, that transmit signals to telephone handsets. The first and second generations (1G and 2G) offer quite slow and unreliable connectivity, often 14 Kbps (kilobits per second), slower than most desktop computer modems, which operate at 56 Kbps. CDMA (code division multiple access), TDMA (time division multiple access), and GSM (Global System for Mobile Communication) all rely on connection-based data transmission rather than packet switching. In other words, the data connections in these standards work like voice calls and are priced the same as for voice calls in most cases. GPRS (general packet radio service) is an emerging, packet-switched 2.5G standard from the GSM, promising faster connect times and offering the possibility of more attractive pricing plans for data connectivity. Packet-switched networking enables data connectivity to be 'always on', and for carriers to charge by the amount of data transferred rather than the amount of time spent on a data call. GPRS is part of the GSM's M-Services Initiative, which combines packet-switched technology, graphical user interface Web browsing, and fun features such as ring tones, games, and wallpaper downloads to mobile handsets.

Conclusions

The industry is changing: devices are getting smaller and lighter but at the same time are becoming more powerful. Phones are sprouting keyboards, cameras, and MP3 players. Designing for these devices is becoming increasingly difficult, as each manufacturer is developing unique user interfaces, even when they use third-party operating systems, such as Symbian OS. However, excellent user interface guidelines are available from manufacturers that describe the appearance and behavior quirks unique to each handset. The design process outlined in this book was developed for all handheld electronic devices. This process, combined with the manufacturer's guidelines, will provide you with ample information to design applications.

Summary

There are three essential types of handheld devices: mobile telephone handsets, pagers, and PDAs. Though many newer devices blur this distinction, design principles will be dictated by one or more of these formats. All handheld devices share two common challenges: small display and problematic data input. Handheld phones generally support voice communication, contact management, and data connectivity via WAP or i-mode. Data are entered on a standard 12-key keypad with the help of various softkeys. PDAs customarily support address books, calendars, to-do lists, email management, and note taking. Data are entered via styli or through desktop synchronization. Pagers support one-way or two-way email and now often feature QWERTY keypads. Distinct rules governing the design of labeled buttons, softkeys, silk-screened buttons, keypads, roller wheels, rocker controls, and touch screens all aid in promoting optimum usability. Though various data entry systems have their pros and cons, no system has yet been developed that rivals the efficiency and ease of use of the standard desktop keyboard and mouse.

Handheld displays are also problematic. Telephones in particular often have severely confined display areas and, when working with them, one must design applications with the most restrictive cases in mind. Displays should be limited to 4 × 12 characters, including soft keys, and blank lines should never be used as they confuse users. Expansion options (hooking up to cameras, scanners, etc.) and connectivity (transferring information between devices) provide further usability challenges. The currently balkanized world of competing connectivity standards and incompatible operating systems precludes organized and rational development of industry-wide norms. In particular, WAP, the wireless application protocol, has suffered a rocky debut for reasons as diverse as poor implementation and questionable marketing decisions. Despite its stormy early tenure, however, WAP appears to be here to stay.

3

Information Architecture: Process

Overview

'Information architecture' is a relatively new term that was coined in 1995 by Richard Saul Wurman in his book *Information Architects*. He defines the term as the visual organization of information. Graphic design is the visual treatment of information. Together, information architecture and graphic design compose user interface design. User interface design is the design of interaction methods that provide access to information, including the graphic treatment, an essential element of any user interface.

I was first introduced to the practice of 'human interface design' when I worked as a summer intern at Apple Computer in 1989. Apple's original *Human Interface Guidelines: The Apple Desktop Interface* defined the human interface as 'the sum of all communication between the computer and the user'. Apple consciously rejected the word 'user' in favor of 'human' in 'human interface'. Users are humans, after all.

Some would argue that 'information architecture' is simply a more appealing term meaning the same thing as 'user interface design'. In this chapter, I use both terms interchangeably, but not to discredit the essential contributions of graphic designers, whose great work has made the desktop Web more attractive and easier to use.

Information architecture for handheld devices is an emerging field. This chapter presents the process of designing information architecture specifically for handheld devices, addressing each of the aspects of the discipline from that point of view.

Relationship between Information Architecture and Other Disciplines

Information architects need to work and play well with all the departments of a product company. Marketers come up with the ideas. Information architects map these ideas out with schematics and flowcharts. Graphic designers then put an attractive face on the ideas. Developers breathe life into them. Throughout the process, usability specialists evaluate the ideas, prototypes, and product to identify ease-of-use issues.

Who does the information architecture work when there is no dedicated information architect? The answer is obviously 'the one who is best suited to it'. Developers, marketers, project managers, quality assurance engineers – all of these people have the capacity to learn and practice information architecture. The relationship between the information architecture (IA) function and the other disciplines is still worth exploring, though.

Marketing

Marketing drives the IA process by coming up with ingenious ideas. Information architecture takes those ideas and comes up with product concepts. An 'ingenious idea' might be for a restaurant finder application for mobile telephones. Information architects ask questions about this application, such as:

- What phones or other devices will the application support?
- What capabilities does the application require?

Marketing may make a business case for the application, as follows:

- The application must generate US$1,000,000 per year.
- The application must support advertising.
- The application must gather 50,000 page views per day.
- The application must gather names and addresses of users, so that mailing lists can be sold.

The information architect would then design an architecture that supports revenue-gathering by microtransaction (pennies per page view), macrotransaction (monthly dues), advertising, membership fees, etc. Registration or some other form of gathering user data would be required. Marketing provides the ideas and information architecture provides the means.

Graphic Design

While marketing precedes the IA process, information architecture can precede or work concurrently with graphic design. Graphic design is equally critical to a product's ease of use as information architecture, but graphic design starts with information architecture as its framework much like an interior designer requires architectural blueprints before he or she can do the job of selecting furniture, fabrics, colors, and floor coverings. The information architect lays out the pages; the graphic designer draws the icons, specifies the type styles, colors, patterns, and graphic accents. Information architecture usually comes before graphic design, but the two disciplines need to work together to be most effective.

Engineering

Information architecture is related to engineering in the same way that architecture is related to structural engineering: after an architect drafts a three-dimensional layout, the engineer takes these specifications and selects the materials necessary to support that layout. Alternatively, he or she may indicate when a structure is unsound or, in fact, impossible to realize. Developers often 'push back' on IA designs, saying they are impossible. Unlike architecture, nothing in software is usually impossible, but when timelines and available resources are considered, it seems as if *most everything* is unworkable. The IA–developer relationship is more successful when the information architect has a fundamental background in computer science and can defend designs based on informed knowledge.

The Design Process

I have just described how different disciplines work together to produce information architecture. The flow diagram in Illustration 3.1 describes the process for product design and development.

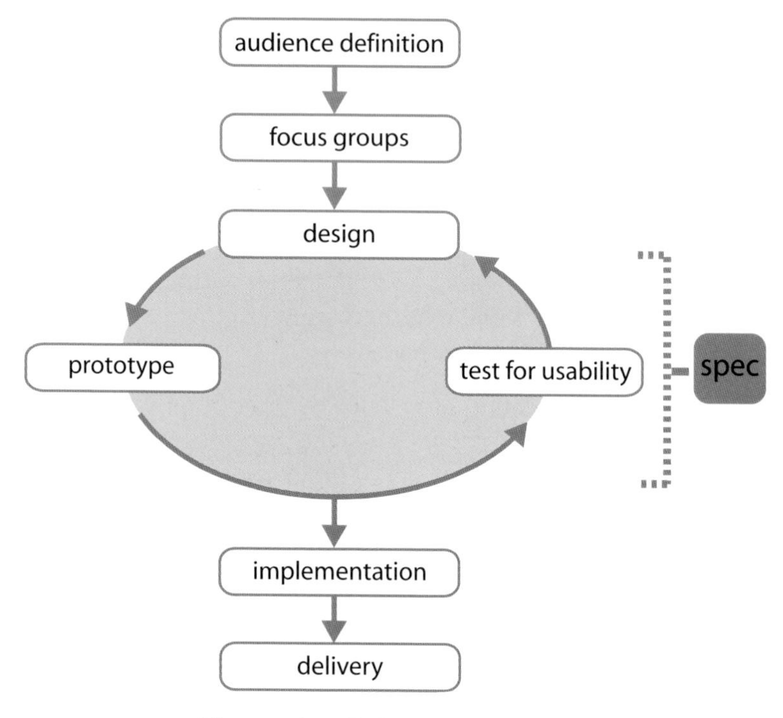

Illustration 3.1 The design cycle

Audience Definition

The first step in any user interface design project, wireless or otherwise, is to define the product and its target audience. Audience definition is covered in significant detail in the section on Audience Definition below (pages 72–80).

Focus Groups

Sharing your product idea with target members of your audience will provide you with a jolt of reality. Knowing how people respond to your product will give your team information about how to proceed with the design or let you know whether the concept needs to be revisited. Focus groups, a traditional form of market research, involve gathering groups of people and interviewing them together. This type of research is qualitative and opinion-oriented. Different focus groups can have very different overall opinions. I recommend a minimum of three focus groups per concept tested per geographic region.

Although I discuss usability testing at length in the chapter on *Usability Testing* (Chapter 6), I should mention that a key distinction between it and focus groups is that usability testing measures respondents' performance as they attempt to complete tasks, whereas focus groups merely ask respondents their opinions.

Design Cycle

By 'design', I mean both the information architecture as well as the graphic design of a product. Most project managers overlook the iterative nature of design, unfortunately. It is important to allocate sufficient time for design, and design by nature is an iterative process. Most teams iterate on designs based on team comments and critique. I recommend at least three iterations of the design, with usability testing between iterations.

Implementation

The implementation phase is when developers code the application and when quality assurance testers identify bugs. Developers always have a lot of detailed questions for information architects during the implementation phase, since it is nearly impossible to document every aspect of a user interface before it is built. Sometimes quick usability tests are required to resolve specific issues.

Delivery

Software developers produce candidates for release to the public near the end of the development process. These release candidates are initially referred to as **alphas**, which are typically 'buggy' and prone to crashes. **Betas** come next, which look and work much like the final release but still may contain many bugs. Delivery is when the application is released to customers, either in the form of a prerelease alpha or beta, or the final 'golden master'. Released products can be referred to interchangeably as 'shipping products', 'final products', 'customer-available', or 'golden masters'.

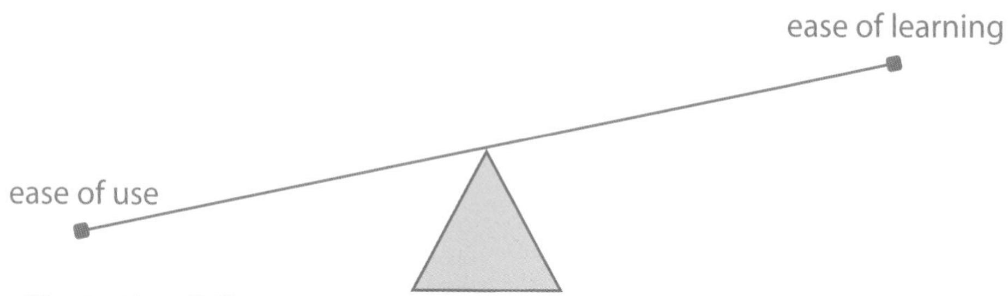

Illustration 3.2 The balance of ease of use compared with ease of learning: examples include the 'End' button as easy to use, and the Weather Channel on Sprint PCS® as easy to learn

Ease of Use and Ease of Learning

Ease of use is the quickness of completing a task. Ease of learning is the intuitiveness of completing a task. You must strike a balance between the two (see Illustration 3.2).

Text entry using T9®, AOL Mobile's predictive text technology (see the section *Triple Tap and T9®* in the chapter on Handheld Devices: Chapter 2, page 44), is very easy to use, but very difficult to learn. Text entry using triple tap is very easy to learn, but very difficult to use.

The 'End' button is an easy-to-use method for terminating a wireless Web connection on the NeoPoint 1000 phone. However, it proved difficult to learn for respondents in a usability study that Usable Products Company conducted in late 2000. Most respondents looked for an 'Exit Wireless Web' function on the main menu but could not find one and failed the task. See Appendix B, on the Sprint PCS/NeoPoint 1000 Usability Study for the full text of this study's findings.

Checking the weather with most services available form the Sprint PCS® Wireless Web was easy for respondents to learn but required many tedious steps in order to specify the city for which the weather was given. In contrast, checking the weather using The Weather Channel from Sprint PCS® involves the following steps:

- Open The Weather Channel (link from Sprint PCS® portal)
- Select 'Cities'.
- Toggle from T9 to NUM; enter ZIP Code (e.g. '10014').
- Select 'Weather'.

- Select 'Current Weather'.

Voilá! The weather comes up in five steps. Easy to learn, hard to use.

Suitability of Applications

Before you design an application, think about whether or not it is even appropriate. Think of the nature of the device the application's communication needs, and the speed and reliability of the intended device's communication ability when designing the application. Desktop Web applications translate poorly to all handheld devices because the desktop Web is optimized for fast processors, large screens, and a captive audience. Training applications are also terrific for the desktop, but terrible for any handheld device. In contrast, a troubleshooting application is ideal for a handheld, since it can be taken to where the problem is located, providing advice when and where it is needed most.

Applications for the Palm OS® and Pocket PC are being deployed in hospitals at a dizzying pace. Physicians are writing prescriptions, orderlies are receiving instructions, pharmacists are scanning medication bottles, and nurses are recording patient information on handheld computing devices. Dozens of applications are available, and hundreds are on the horizon. None of these applications is especially suited to the desktop, which is not portable.

However, note-taking on a handheld is terribly ineffective. Writing on the tiny screen is far less satisfying than writing on paper, and handwriting recognition is very poor. So the pocket word processor will not be a runaway success, but a *structured* note-taking product, such as a patient record, would be far more likely to succeed. A questionnaire with check boxes and radio buttons with few type-in forms is more successfully deployed on a handheld than is a series of fill-ins.

As we have seen, wireless devices are for people who are in motion. The recommendations made in the next section are a great starting point for developing your own wireless applications.

User Interface Design Guidelines for Handheld Devices

Design for Users on the Go

Whether in the back seat of a taxi or walking down the street, people are likely to need their handhelds to perform in distracting situations. In restaurants, meetings, or elevators, handheld users are likely to be conversing in a hurry, so designs must include **context** and **forgiveness**. While desktops accommodate 'surfing' – users meandering down the byways of the Information Superhighway for hours at a time – wireless devices are more about instantaneous search and retrieval (see Table 3.1). The stakes for wireless usage are higher than for desktop: per-minute charges apply, data transfer is slower, and connectivity is sketchy. Wireless users may be using their leisure time to gather information, but they typically have immediate goals.

Table 3.1 Comparison of use of desktop and wireless Web

Desktop Web	Wireless Web
Comparing prices of flights and making reservations	Checking status of a particular flight
Gathering background on a company, including maps	Getting driving directions to a company – while on the road
Researching a medical condition	Monitoring a medical condition
Reading a movie review and/or watching a trailer	Purchasing a theater ticket to avoid the line
Analyzing a portfolio of stocks	Placing a trade
Checking a product's availability	Scanning in warehouse inventory

'Select' vs. 'Type'

Text entry is often required, but typing on a handheld device is extremely difficult unless a keyboard is attached. Even when a device supports a keyboard, though, your user may not always be in situations where it is feasible to attach one. Alternative input methods such as Graffiti® handwriting recognition software and on-screen keyboards are painfully slow for data entry, so, when possible, offer a selection mechanism rather than requiring typing.

The exception proves the rule: when the number of choices is too high for easy selection, such as for stock quotes, text entry may be more appropriate. Typing in 'IBM' is easier than choosing among industry sectors, then selecting the tickers that start with 'I', then scrolling among the dozens of companies that are returned. Of course, in a user interface that relies on 'select' over typing there are many options for how to organize the data and each user may have a different perspective. In my experience, the balance between a selection-based and a typing-based user interface changes when the number of choices is reduced to fewer than 20, especially if the choices can be further broken into categories, such as for cities within regions within countries.

Do not take this guideline to be absolute, or you may hurt your audience more than help them. Just keep in mind that users will have difficulty when entering data, and design for them accordingly.

Be Consistent

Borrow from well-designed applications when user interface standards and guidelines are either not available or are not yet developed enough to support your interaction mechanism. However, do not invent new user interfaces when one of the existing interfaces will do nicely. Your users do not want to learn new techniques to access information unless the new techniques will save them a great deal of time and effort. Think of how time savers such as the fax machine and email have quickly reduced the amount of hand-delivered mail.

Use the same terminology and interaction schema within the same application and between applications, which will reduce the learning curve for new features.

Consistency between Platforms

Many of you will be reading this book as a prelude to porting an application's user interface from Windows or the desktop Web to WAP or a Web-clipped site. While consistency can be an effective tool to increase ease of use, it can hinder good design when migrating a user interface from platform to platform. My immediate advice would be to retain terminology and processes only when they are equally appropriate for the smaller screen. When either becomes an obvious hindrance in the wireless environment, you will need to redesign from the wireless user's perspective.

Imply User Control

Provide the *illusion* that the user is in control. We all know that the application is really the controlling entity, but no one wants to feel controlled by technology. We want to feel enabled by technology, not trapped by it.

The desktop user interface gives users the illusion of being able to manipulate objects by clicking and dragging them. Whereas 'switching applications' was as easy as clicking another icon, WAP takes us forward and backward within lists and lists of menu options. Instead of being able to locate an application graphically and click it, our interaction has been reduced to hunting through preorganized menu lists to locate what we want. However, menus can be designed to provide cross-linking between areas of applications and even between entire applications, giving users freedom similar to what they left behind on the desktop.

The clipboard model is the ability to cut or copy data and paste it elsewhere. It gave desktop users a great deal of control. This model, which made the desktop user interface so powerful, is usually not available on wireless devices. Even calculators supported some concept of a clipboard with the 'M' key, but WAP, phones, and pagers neglect it completely. The Palm OS platform supports it to a degree but hides it behind a menu that is difficult for many users to access.

You can compensate for the missing clipboard model by enabling common actions on visible data. For example, enable users to 'call' when they have selected a number that correlates to your country's dialing system. Provide access to email when users select text that includes an '@' symbol. Anticipate how a user will act on data and build his or her desired actions into your design.

Design Stability

Since wireless data connections are prone to failure and will remain so for the foreseeable future, it is critical that your applications provide a stable user interface. By 'stable', I mean that when the network connection drops the application should restore state and context once the network goes back online. For example, if a user is entering parameters for a flight reservation, and the network goes down mid-process, once the network connection is restored the process should continue without requiring reentry. The network was unstable, but your design need not be. You could further increase the user's confidence by displaying a message such as 'Parameters restored' (see

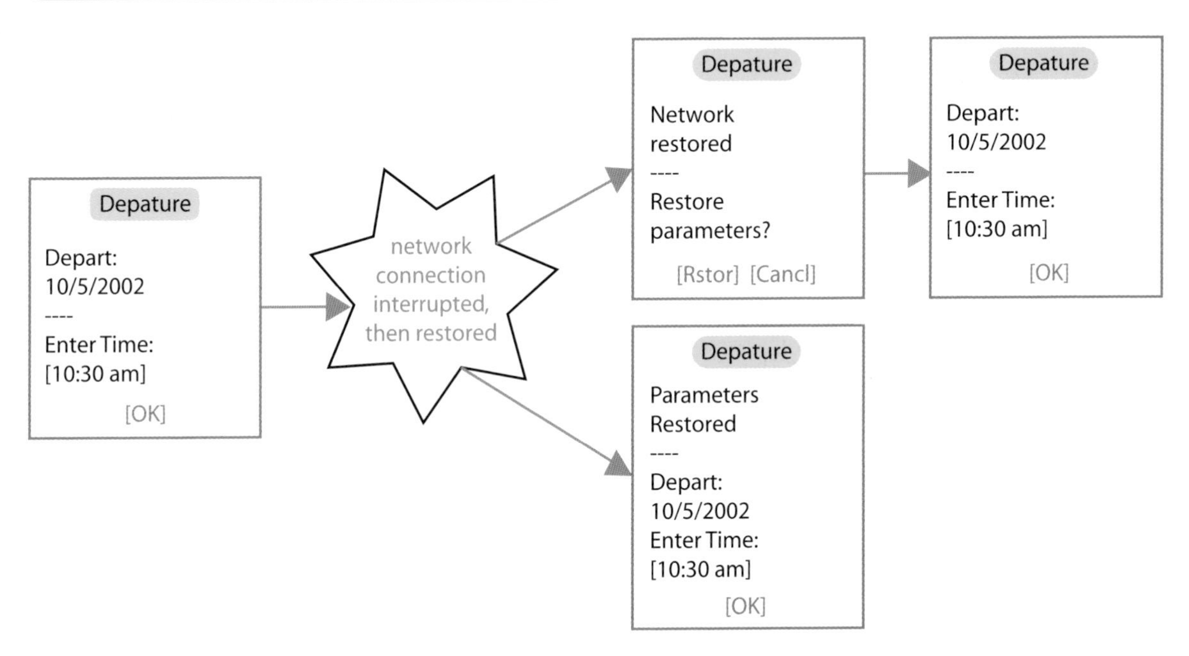

Illustration 3.3 Anticipate user concerns when network stability is in question. Instil confidence by keeping the user in the loop and retaining state. Either solution is acceptable

Illustration 3.3). Offering to restart the process could be helpful, but if so the data from earlier steps should prepopulate entry fields to the extent possible rather than displaying empty fields on data entry pages.

Provide Feedback

Each page of an application should provide the user with enough information to understand what the application is and how to navigate from that page. If the display is too small to provide a detailed explanation, a clear path to that information must be available, either from a softkey menu or the device's 'Menu' key.

Forgiveness

If a user makes a mistake, the user interface must offer a means to correct it, even if that means just the phone's 'Back' button. Alternatively, the 'Menu' key could support 'Undo'. The richer your 'Undo' functionality, the more forgiveness your application provides. Also, multiple-level undo is preferable to single-level undo.

Use Metaphors

When designing applications, use metaphors from the real world. Instead of referring to a place for storing information as a 'repository', you can call it a 'file cabinet' or a 'folder'. The 'desktop' metaphor would be a poor choice for handhelds and phones, since the display is too small for the user to make the conceptual leap. The 'bookmark' is a very effective metaphor for the wireless Web, especially since entering URLs into a handheld is so awkward.

Clickable Graphics should Look Clickable

The converse is also true: images that are static and not linked should not appear clickable. 'Clickable-looking' means that they should have defined borders and/or should have high contrast with the background color.

Use Icons to Clarify Concepts

On tiny displays with no graphics support, this recommendation is mostly irrelevant, but on color, bitmapped displays, icons provide users with additional assistance. Be careful to design your icons as representations of concepts, not as 'hieroglyphics' (hieroglyphics are pictograms that combine thoughts by overlapping images from the real world). The best icons are very simple representations, usually nouns. Verbs are difficult to convey as icons, such as 'connect' and 'disconnect'. Whereas verbs are more abstract, nouns are concrete. Further, technology-centric nouns are tough to convey as icons, such as 'Menu'. The obvious US metaphor, a dining menu, is a poor choice.

Icons may not be immediately obvious to users, but they can be memorable. A 'triangle' image that displays a menu would not immediately be recognizable as such, but after its first use it becomes a memorable image to the user. In the user's mind, 'the triangle button brings up the menu'. There are many icons that are part of the vernacular: use these for their intended meanings instead of designing your own. For example, the envelope is an ambiguous icon, since it could mean 'email' and it could mean 'voicemail'. A good compromise would be to consolidate both email and voicemail in the same location.

Information Architecture Process

The process of designing information architecture must be initiated with requirements that come from marketing. Information architecture can be for new products and applications, new releases of an existing application, or porting an application user interface from desktop Web to wireless.

The first and most important step in information architecture is understanding your audience. We consolidate our grasp of our product's market in the section on Audience Definition (pages 72–80). Marketing provides requirements, presumably with the audience in mind. However, the marketing people may not have a detailed audience definition, so you might as well start with an exercise of identifying your users' common traits.

The next step is to describe the product and its features in natural language. This step is called **scenario development**. Scenarios are in paragraph form and are more easily discussed in meetings and edited or rearranged in flowchart form. The flowcharts, when combined and clarified, are referred to as application maps. I lump 'site maps' into the more general term 'application'. The application maps include all of the individual pages, which are articulated with page maps. The stepwise process is as follows:

- Audience definition: audience definitions are descriptions of who will use the product once it is completed. Users are described by means of their traits and the acceptable range for each trait.

- Scenario development: a scenario is an outline or a model of an expected sequence of events. Scenarios are used in information architecture to capture how a product will be used, from the user's perspective. In this case, a scenario describes a product user's needs and the steps he or she will follow in order to fulfill those needs.

- Flowcharting: flowcharts take the scenarios and break them into functional components, which may correspond to pages in an application, organized from the point of view of the scenario.

- Application mapping: application maps give a quick overview of how the different functional components interrelate, corresponding to the application's navigation and showing each of the pages or functional areas. If the functional areas are represented, further detail will be necessary, provided in iterative scale.

- Page mapping: page maps articulate the individual page designs, together with descriptions of each of the user interface elements.

All told, these components constitute a 'functional specification'.

Marketing Requirements

Marketing requirements should be specified before information architecture, and so I will articulate the requirements for our example application, weather checker.

- The application will provide detailed weather information for all cities, airports, and ZIP or postal codes
- The application will be geographically aware
- The application will support 'weather alerts'
- The application is for WAP devices
- The application supports bookmarking of 'favorite' cities

Audience Definition

Audience definitions are critical throughout the design and development process. Before we can design a Weather Checker we must, then, define our audience. The traits and their ranges are critical for recruiting for focus groups and usability tests, but they are also invaluable for participatory design sessions and other IA activities. Audiences are defined descriptively and by ranges of traits. One or more of the following traits could describe your audience:

- early adopter
- novice and savvy
- Internet savvy

- teenager or senior
- phone, PDA, and/or pager savvy
- everyday consumer
- physically challenged
- complex application user
- online transaction user
- credit user
- enthusiast
- commuter
- high-income and high net worth
- geo-specific

Traits can be described or they can be screened. Describing a trait explains what it is like. Screening for a trait involves writing questions that will identify those for whom the trait is evident and also eliminate those for whom the trait is absent or inapplicable. In the audience definition, traits are described. The respondent screener, in contrast, is a questionnaire designed mainly in order to screen for the traits described in the audience definition. In the following sections I provide brief descriptions of the traits identified above.

Audience Descriptions

Early adopter

An early adopter is someone who might acquire technology for the sake of trying something new. Their possessions and how frequently they use them define 'early adopters'. Some businesses try new technologies in order to locate efficiency improvements. The business case is harder to recruit for unless you have lists of prescreened candidates. In an audience definition, you can describe someone's willingness to learn a new technology although, for recruiting respondents, 'willingness' is hard to gauge.

Novice and savvy

For our purposes, the distinctions between novice and savvy may be hard to distinguish. Someone is a novice if they have used a technology or product for a limited amount of time *or* if they have not become very proficient with the technology or product. Someone who is savvy with a particular technology may only be savvy with one version of it, and a novice – or completely inexperienced – regarding the latest upgrade.

Internet savvy

Defining one's level of Internet savvy depends on the project being considered. I have successfully used the following criteria:

- time spent on the Web per week (including or not including email, school work, etc.)
- ability to define specific terms, such as 'home page', 'search engine', and 'link' in common language

It is not important for respondents to be able to define technical terms; just terms that describe common tools. For example, someone who is Internet savvy may understand what a home page is but not know how to set the default home page in his or her browser.

When recruiting for desktop Web usability tests, we always require respondents to define the 'Back' button successfully. I have found that anyone who cannot describe such basic functionality is not suitable for our clients' research. For wireless browsing, the 'Back' button is still relevant, but less so on many smaller–format telephones. We have not yet established a reliable baseline measure for wireless Internet savvy, partly because of the inconsistency between wireless Web implementations. Our working metric is the number of minutes per month spent on the wireless Web.

Teenager or senior

While most teams suspect the age range to be intuitively obvious, much discussion ensues when actual numbers are required. Teenagers are technically 13–19 years of age, but defining 'young adults' is project-specific. A 'senior' may be 65 or older, or 65–79. Teenagers are likely to pick up new technology quickly. Seniors always need larger type and graphics because of vision challenges. Seniors are also likely to have difficulty with small screen targets and small physical buttons. Older people are also less adaptable to new user interaction mechanisms.

Phone, PDA, and/or pager savvy

All technology device-savvy individuals will understand the battery life of their device. Novices may not be able to state the battery life with confidence.

Someone familiar with a mobile telephone will understand the concept of call interruption, roaming, and may have several friends programmed into the address book. Most phones support an address book, and savvy phone users often take advantage of this popular feature.

PDA-savvy people understand 'hot synching', 'cradles', and know what a stylus is. They may know how to use Graffiti® input methods but they may not know what it is called. Pager-savvy people will know how many lines are on their readout, whether or not their pager works when they leave the region, and will usually receive at least five pages per week.

Everyday consumer

There really is no such thing as an 'everyday consumer'. However, I am frequently requested to locate this elusive audience type for client research. I combine the following traits, with ranges specified on a project basis:

- amount spent online for purchases in the past 30, 60, or 90 days
- amount spent in brick-and-mortar stores in the past 30, 60, or 90 days
- number of stores frequented each week
- print catalogs browsed per week
- Web-retailers visited per week
- income, individual or household
- age range

Physically challenged

Vision, hearing, limb use, breathing ability, or mental ability can all be 'challenged'. Many nations have accessibility requirements for the Web. Table 3.2 lists links to sites with information about designing accessible products.

Table 3.2 Websites carrying information on accessibility for the physically or mentally challenged

Site	Description
www.access-board.gov	The Access Board, a US agency devoted to accessible design. See Section 508, Electronic & Info Technology.
www.w3.org	World Wide Web Consortium Web Content Accessibility Guidelines
www.microsoft.com/enable	Microsoft's recommendations for making accessible software using their products.
www.accessibility.com.au	Australian accessibility information
www.tag.gb.com	A UK site describing disability access in the United Kingdom. Information is primarily about locations rather than electronic devices.

Complex application user

A person who does his or her banking online is a complex application user. Someone who uses desktop software such as spreadsheets and word processors is familiar with complex applications. So is someone who knows how to program names into his or her telephone address book.

Online transaction user

People who buy things online are transaction users. Frequency and money amount are both effective at gauging the level of transaction use:

- number of stock trades (or percentage) completed online, as opposed to trades via an automated telephone interface or through a human broker
- number of movie tickets purchased online
- number of online banking transactions completed

Frequencies are usually specified 'per month' rather than 'per week' or 'per year'.

Credit user

One who is comfortable purchasing products with credit, not necessarily online. Credit users can be identified by the number of credit cards they carry, the type of credit and charge cards they use every month, by the amount added to their balance each month, or the amount of balance they carry from month to month. Credit savvy could be determined by knowledge of interest rates and grace periods.

Enthusiast

Simply put, enthusiasts like certain things, such as sports, technology, automobiles, dogs, etc. Describing an enthusiast is easy, but screening for enthusiasts requires a stretch:

- Which of the following magazines do you subscribe to?
- How often do you play bridge?
- Do you own golf clubs?

Commuter

Commuting can be by automobile, bicycle, on foot, by public transit, etc. The means distance, frequency, and number of passengers are all relevant for this category.

High-income and high net worth

Income can be determined by individual or by household and is usually measured annually. Net worth, the amount of assets a person has, may be a better gauge for your project but is a more difficult trait to measure. In order to determine net worth you must specify whether real estate, retirement plans, and other asset classes are considered in the calculation.

Geo-specific

This user type could be someone who lives in a particular neighborhood, city, state, or country. It could also describe an urban dweller, suburbanite, or rural dweller. Finally, your product may require someone who lives in a multiunit building or someone who lives in a single-family home. All of these possibilities may be relevant, but usually not more than one at a time.

Additional Traits to Consider

Not all of the following traits will be required in order to define your users effectively. However, they should at least be considered before you decide to eliminate them from your audience definition.

Easy Traits

Gender distribution Some products are more focused on women, some on men. Men and women frequently perform similarly in usability, but their interests often are not identical.

Race I have not observed any difference in usability performance based on race alone but, again, different racial groups will have different product and content interests. If you use a recruiting service they will address this issue upon request. If you are recruiting for your own research, define a matrix of ethnicities that is specific to your product.

Household income Who is 'rich' is defined differently by everyone. Note that household income is different from individual income, which may be an unrelated trait. I never use both traits.

Residence location City people have very different interests from country people. Postal codes and telephone area codes are sometimes more effective than city names for determining whether an individual matches your residential requirements, as cities can be quite large. Postal codes can help you target your audience to within a few blocks, if you so choose.

Education level Five years of usability testing data have given us confidence in saying that education level is directly related to usability performance. Yes, there are people who are brilliant but who don't have a high-school degree, but they are the exception rather than the rule.

Job and employer What people do and whom they work for is usually important. Most clients want to screen out individuals who work for their company – or their competitors. Sometimes clients want to screen out journalists, to prevent unexpected press coverage. Screening out certain job types can also reduce the chance that an 'expert' accidentally gets recruited. Recruiting a financial planner for research about a stock trading site for novices would not be appropriate, for example.

Tougher Traits

The traits I classify as 'tougher' are harder to define only because they are more vague. They can be evaluated on several axes, and sometimes require more than one criteria to cover them.

Net worth The amount of money that someone is worth can include any of the following:

- real estate owned (including or excluding primary residence)
- cash, CDs, money markets, checking and savings accounts
- stocks, bonds, and mutual funds
- retirement accounts

Each product that requires net worth as part of the target audience definition needs to identify which of these criteria are relevant.

Cellular phone experience Experience with phones can be evaluated on the following axes:

- time spent on wireless voice calls per week
- time spent on wireless data calls per week
- average monthly wireless phone bill
- number of months having used the wireless Internet
- number of years using cell phones
- hands-free phone use
- number of phone numbers in wireless phone's directory
- frequency of synchronization with PC data

Your team will need to decide which are the most relevant criteria. Sometimes the criteria have an indirect relationship, such as hands-free use. Other times the criteria help determine how avid the audience might be, such as the number of wireless minutes used per month.

Scenario Development

When designing user testing scenarios, begin by documenting ways in which the product might be used. Let's start with an example scenario.

> Joe's business trip: Joe is on the train going from work to home in Silicon Valley. He leaves on a business trip to New York City tomorrow and is curious about the weather there. He uses a WAP phone to check the weather at Kennedy Airport.

In Joe's scenario, I haven't yet described an application. I did, however, describe many of the components that might affect the application's design. We can glean the following from the scenario:

- the platform is a mobile phone.
- The scenario describes a physical location – perhaps the hardware and software are (or can be) geographically aware.
- Relevant parameters are: 'tomorrow' and Kennedy Airport in New York City.

Most of the time you will know for which application you will be developing scenarios. The questions to ask yourself are:

- Why would someone use the application?
- Under what circumstances would someone use the application?
- What are the various things someone would want to do with the application?

Once you have these questions, you can start building scenarios. A scenario is a story in which your application plays the key role. In the scenario above, Joe's need for weather information in another city while he is away from his desktop compel him to use the Weather Checker to get information. Here's another scenario for the Weather Checker:

> Mary's weather mania: Mary is a weather information buff. She always wants

> to know the weather in cities all over the globe. The problem is, she never has time to check during the day while at work, so she looks up cities while at lunch, during her commute, and in the evenings. She has London, Moscow, Tokyo, and Sydney bookmarked for easy access. She keeps track of her local weather by way of alerts that she has sent to her phone, based on her current location.

In Mary's scenario, the approach is coming from another direction:

- Weather Checker is a geographically aware application, but we know that users will not always want to get the weather information where they are located at the time.
- Information alerts in Weather Checker must be supported
- Weather Checker must support bookmarking of favorite cities.

Flowcharts

Flowcharts take a scenario narrative and reduce it to steps. Flowcharts utilize specific shapes to describe different actions, as defined in Illustration 3.4. I use a drafting application to draw flowcharts and other images, but most modern word processors, spreadsheets, and presentation programs offer the basic drawing tools that you will need.

Illustration 3.5 shows the very simple flowchart for Joe's business trip, as described in the previous section. Note that 'enter city information' is a process. The narrative for Joe's business trip might be diagrammed alongside the application map to help readers fully comprehend the situation.

Illustration 3.6 shows the slightly more complicated flow chart for Mary's weather mania, also described in the previous section. I left in the process flow for 'enter city information', but added complexity for the bookmark feature. I did include arrowheads, in order to increase clarity.

There are some designers who have a lofty code for how to create flowcharts. I stick to keeping the information simple. The rules are restricted to using the shapes in the manner described above. If you feel the need to create a new shape, by all means do so.

As you flowchart more and more scenarios, you will naturally want to include more functionality in a single chart. Try not to – save that detail for the application map, which I will present next.

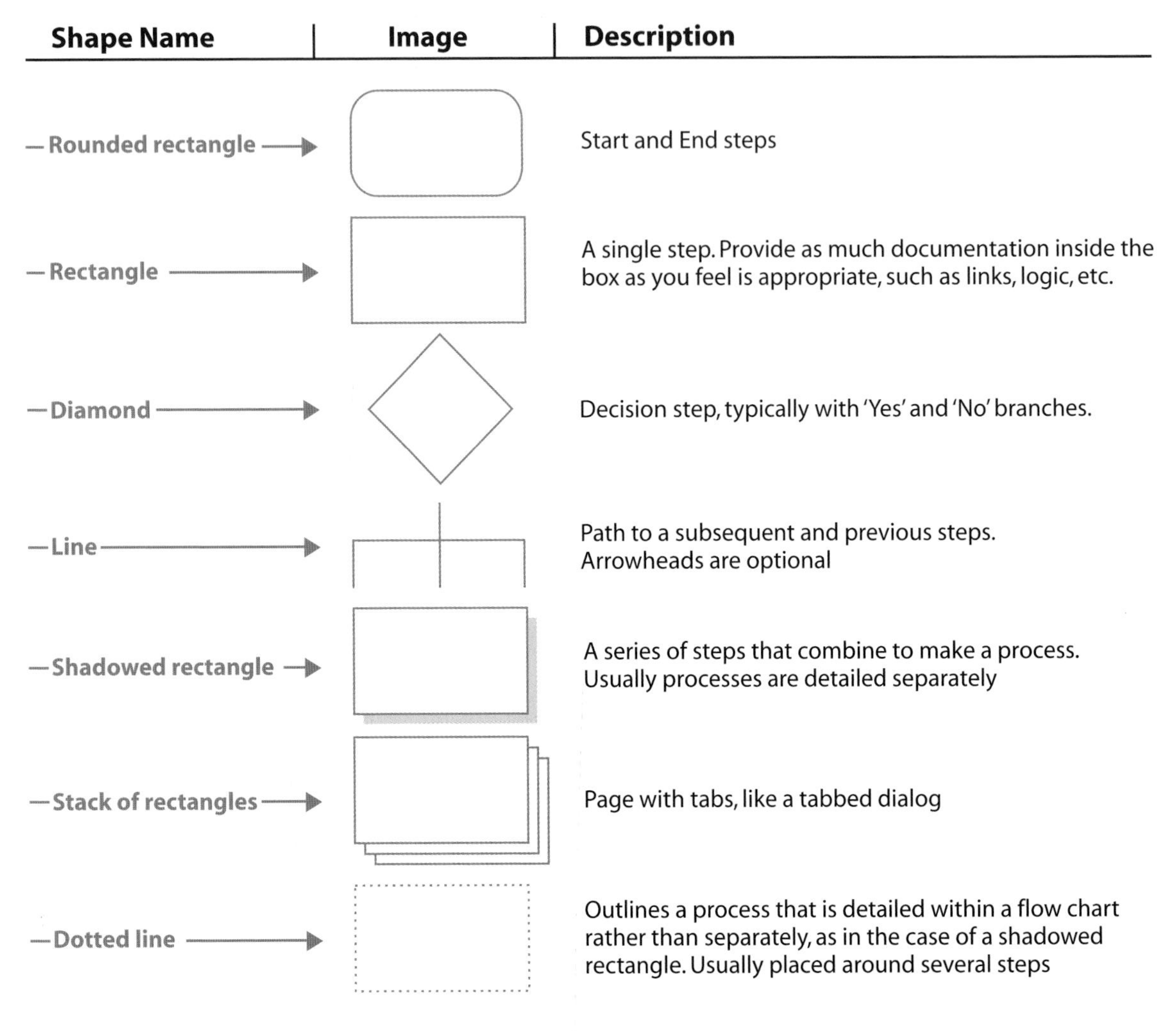

Illustration 3.4 Flowchart shapes

Application Maps

As mentioned earlier, the application map is a far-away view of the application. The application map is where the relationships between flowcharts become apparent. Most of the visual imagery in an application map is the same as in a flowchart, but the focus is different. The flowchart documents a single

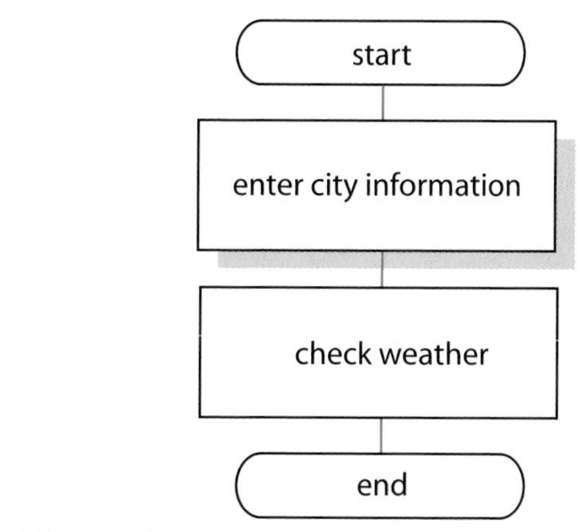

Illustration 3.5 Flowchart for Joe's business trip

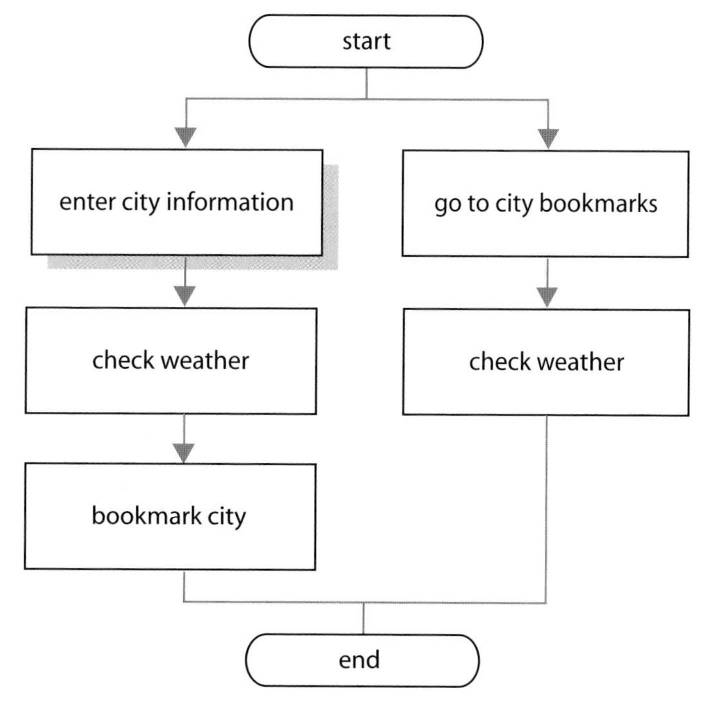

Illustration 3.6 Flowchart for Mary's weather mania

scenario, whereas the application map documents the pages and processes of the application itself. The application's navigation is evident from the lines drawn between boxes.

For larger applications, it may be necessary to provide the application map in scale levels, so that the map can be reasonably documented on standard office paper ($8\frac{1}{2}" \times 11"$ or A4). For the scale levels, use process graphics (shadowed rectangles) or combine functional areas with a dotted-line border.

Illustration 3.7 shows an application map for Weather Checker (WCh). Notice the new visual, the 'home plate' graphic,

It refers to a subroutine label. When depicted point-down, the graphic is the label. When depicted point-up, it is the destination. In the above application map, I deliberately kept the detail level low, which enables us to understand the navigation and flow of information.

Page Maps

Page maps detail the full amount of information conveyed on an application page. Page maps can be documented in a functional specification or within an application map. In Illustration 3.8 example, I have included all of the page maps for Weather Checker in its application map.

Page maps for wireless devices contain the following elements:

- page title: depicted as it will be shown in the final application
- page text and graphics: depicted as it will be shown in the final application
- variables: populated at runtime; represented in angle brackets (<, >)
- links: hyperlinks; represented in square brackets ([,])
- buttons: unavailable in WAP, except for Openwave's most recent Mobile browser; the button label is shown, but the button's destination is shown as a line from the page or an accompanying page description
- softkeys: platform-dependent; represented in square brackets at the bottom of the page

Some platforms support menus, which are depicted in functional specifications but not in page maps since the menus appear after a specific button press.

Conclusions

Understanding who your audience is before you start designing sounds like good common sense, but you would be surprised how few design projects start out that way. The same goes for developing scenarios, flowcharting them, etc. The most important aspect of the design cycle presented in this chapter is the fact that design *is a cycle*, with iterations determined by usability test results. If your goal is to design a great product, defining your audience and designing for that audience will get you most of the way there. In order to achieve the goal of great product design, however, the design must be tested with members of that audience.

Summary

This chapter presented a process for producing information architecture, which is the organization and layout of information as it is gathered from and presented to its audience. It started by defining the concept, then described the discipline and how information architects relate to other software and hardware development professionals. It continued by presenting guidelines for good information architecture for wireless products. These guidelines evolved from desktop software guidelines developed more than 10 years ago by Apple Computer for the Macintosh™.

Good information architecture requires an understanding of whom you are designing for. The chapter explained how an audience is defined. Once you understand your audience, the next step is to describe what your product will do, in the form of scenarios. The scenarios are broken down into defined steps and are documented with flowcharts. The flowcharts are combined into application maps, which provide an overview, and page maps, which provide full detail of each page and how it will look and function.

The next chapter presents the practice, or nuts-and-bolts, of information architecture for wireless devices. It details the elements of hardware and software for PDAs, phones, and pagers and gets into the quirks of each device type. This chapter dealt with fundamentals of information architecture, while discussing issues important for the entire wireless platform. The next chapter deals primarily with the specifics.

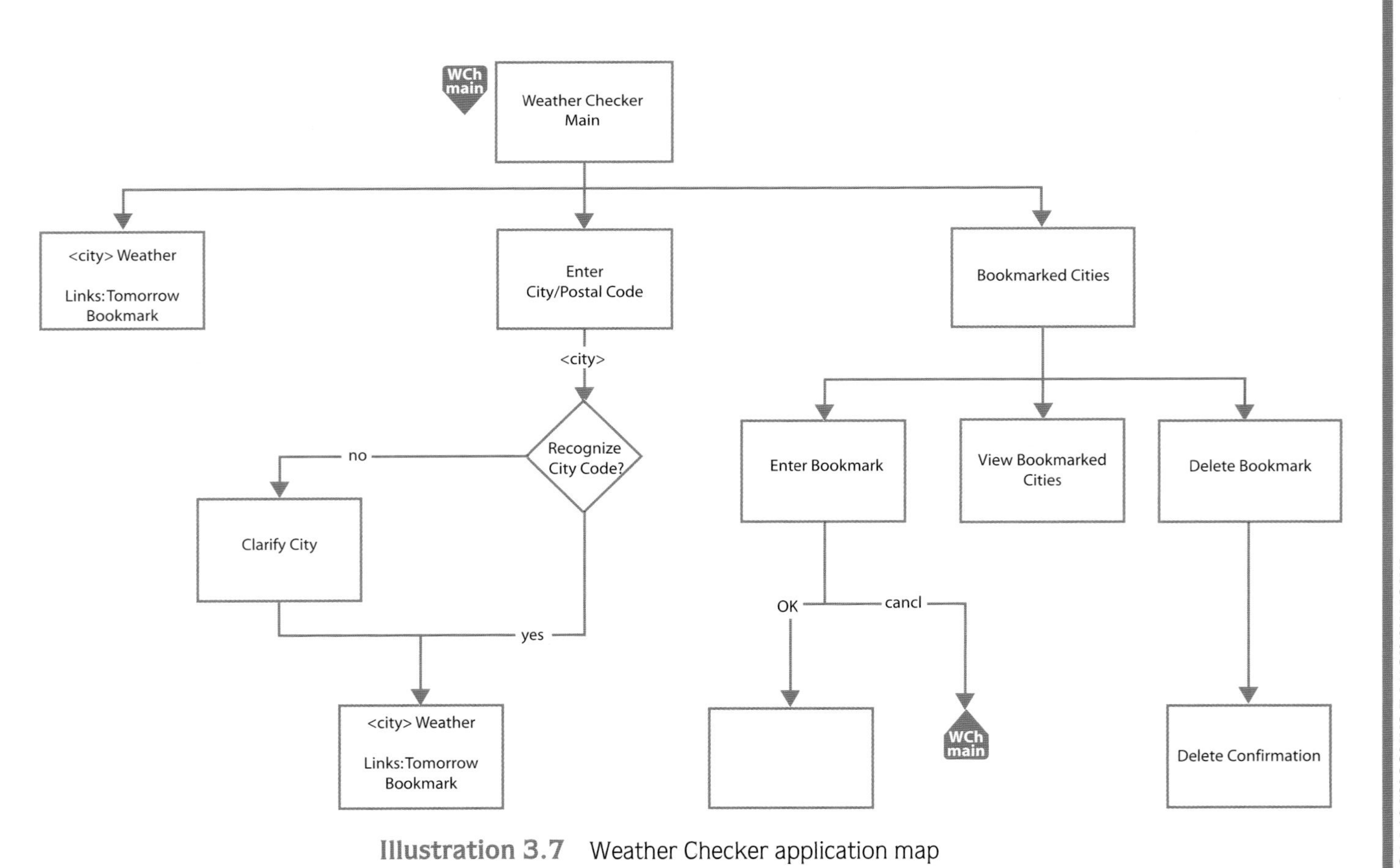

Illustration 3.7 Weather Checker application map

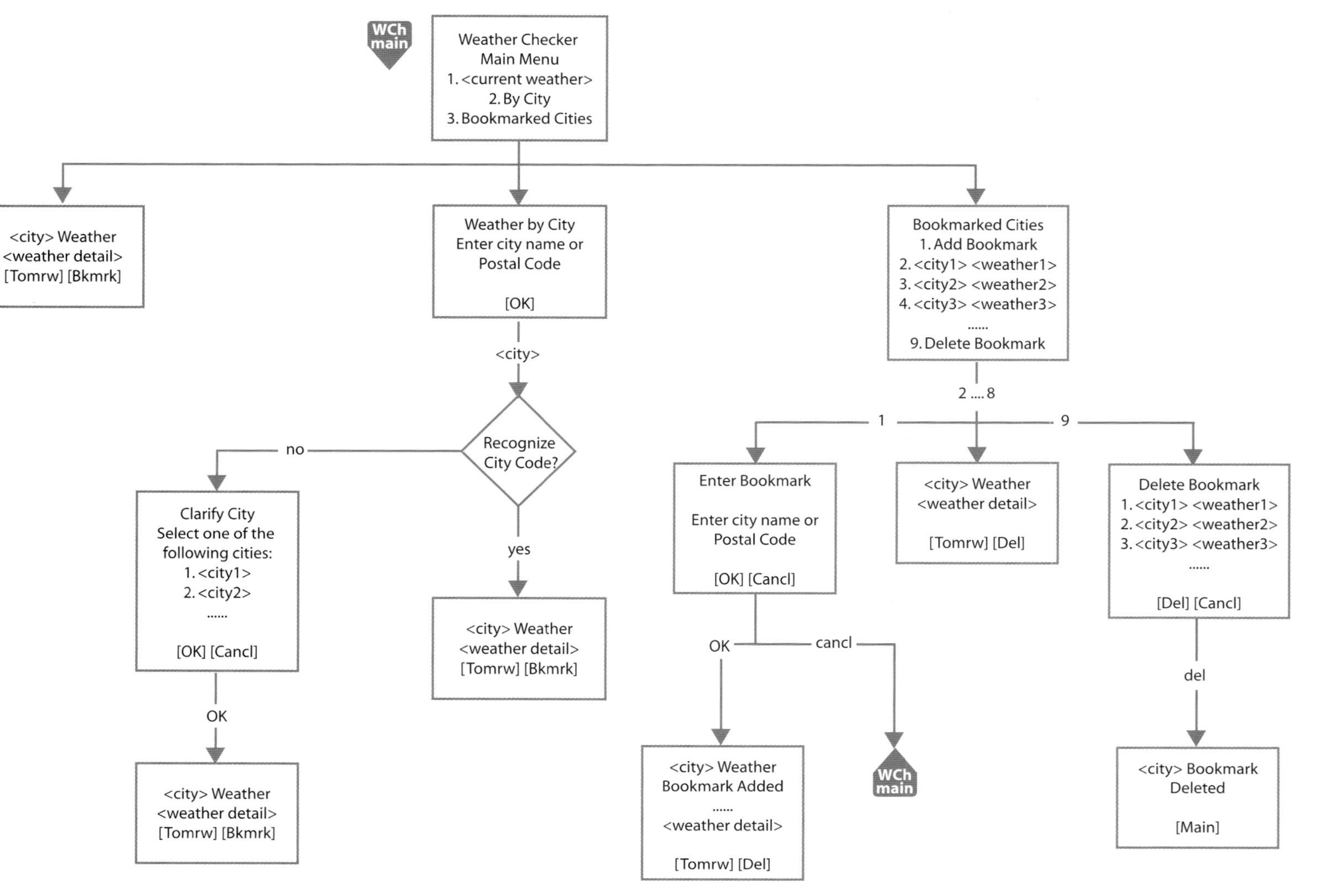

Illustration 3.8 Weather Checker page maps laid out in an application map

Information Architecture: Practice

Overview

The previous chapter presented the process of information architecture: how one designs an application. This chapter is the direct application of the concepts in that chapter – as they relate to handheld devices. First, devices' software user interfaces will be deconstructed to give you a sound understanding of the palette of interaction mechanisms available. By gradually moving on to higher-level user interface constructs you will gain an understanding of how the building blocks come together. Finally, you will find specific advice on how to design for the different handheld platforms: WAP and phones, PDAs, and pagers.

User interfaces of handheld devices are on a continuum of sophistication, starting with character-based WAP, the most limited operating environment. At the other end of the spectrum is the touch screen PDA, which offers precise positioning of graphical user interface controls and drag-and-drop functionality.

Most existing telephone handsets support character-based WAP. PDAs and pagers support graphical user interfaces. Adding to the confusion is the newest release from Openwave, a graphical WAP browser. The graphical Openwave WAP browser, which adheres to the GSM's *M-Services Guidelines* (www.gsmworld.com), offers many of the features of a graphical user interface within the confines of a tiny screen and handset button controls.

Graphical User Interface Controls

Although handhelds come in many sizes and shapes, several user interface constructs are common to graphical user interfaces of all handheld devices. This section steps through each user interface control and provides recommendations for its use.

Icons

Icons are graphical representations. There is no standard for size, color depth, or style among handheld devices. The following are common sense principals.

- Keep icons small, but recognizable. All handheld devices have small displays and although your icon may be beautiful it takes up valuable screen real estate. However, if your users cannot tell what the icon is meant to represent, it detrimentally affects the user interface.

- Use high contrast colors. Handhelds are used in different types of lighting conditions, and in sunlight they can be 'washed out'.

- Make icon edges stand out. Use black if you can. Thick lines are easier to identify in soft light.

- Avoid three-dimensional icons. 3D looks great on the desktop where pixels are in abundance, but it loses its allure on the small screen. Keep your icons flat.

- Hieroglyphics were left behind for a reason: they were hard to read. Keep icons simple, and avoid combining pictures to make a 'sentence'.

- Nouns are easier to 'iconify' than verbs. A paperclip is easier to conceptualize than 'attach'.

- Be consistent when possible. If you have seen an icon for a given function, use it. Coming up with new graphics for old concepts will confuse your users.

- View your icon at different sizes, since the handheld may scale it without your control. Some operating environments allow software to register multiple icon sizes – take advantage of that opportunity. When creating icons at different sizes, make them similar in appearance and style so that users will not be confused.

- Icons can be intuitive or memorable, or both. When they are neither intuitive nor memorable, they are not worthwhile. Try using an unusual color or shape, or incorporate an element of your company's branding into the icon.
- Avoid placing images within rectangular borders. Icons with distinctive shapes are more memorable.
- Do not use text in icons. Bitmapped text is difficult to localize, and the type is likely to be too small to read effectively. Use text *with* your icons if you like, but not *in* your icons.

Audicons

Audicons are the sound equivalent of icons. Brief, distinctive sound effects can add usability and enjoyment to your application. Audicons are an effective way to inform your user of a task's completion or success in an awkward direct manipulation task. Sound in applications should never be the only user interface, as many users will mute their devices.

Menus

All handheld devices support menus of one type or another. Menus are sets of commands that are hidden behind a graphical button, menu bar, or tap-and-hold paradigm. Menu commands affect the entire application or portal that they cover.

Menu bars are an interactive, hierarchical command structure that enables a user to navigate within an application. In the Palm OS® environment, the menu bar is accessed through the 'Menu' hardware key or by tapping the title of the page. On phones, the menu is often accessed through a dedicated hardware key.

Popup Menus

Popup menus appear in place over a call to action. The 'call to action' typically has a pushbutton-like appearance with an arrow or triangle image superim-

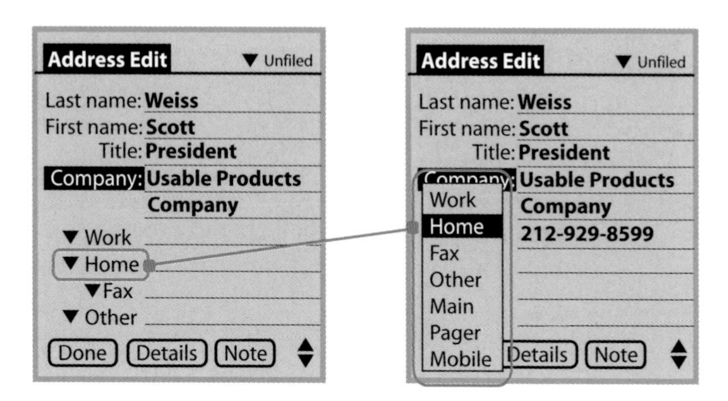

Illustration 4.1 'Home' is clicked to access a popup menu. Reproduced with permission from Palm Inc.

posed (Illustration 4.1). Upon a button press (or stylus tap), the menu appears and enables the user to select an item. Popup menus are often used to identify the content adjacent to them, functioning similar to a compressed set of radio buttons. In Illustration 4.1 the labels adjacent to several text entry fields are popup menus. A popup menu item may launch a dialog box if its label is trailed by three ellipses ('...').

Text Entry Fields

Text entry fields are boxes or dotted lines where text must be entered by the user. They usually have a text label above them or to their left. Most user interfaces indicate when a text entry field is active by way of enhancing the border or showing a blinking insertion point.

Different products support different filtering for text entry fields, such as quantity, numerical, text, or password. Quantity filters beep after a certain number of input characters. Numerical and text filters typically beep and otherwise ignore unacceptable input. Password filters have several implementations, but almost always change character inputs to asterisks ('*') as the insertion point moves forward.

Check Boxes

A check box is an on–off control. Check boxes may appear alone or in groups (Illustration 4.2), but in groups check box functions are unrelated – one check

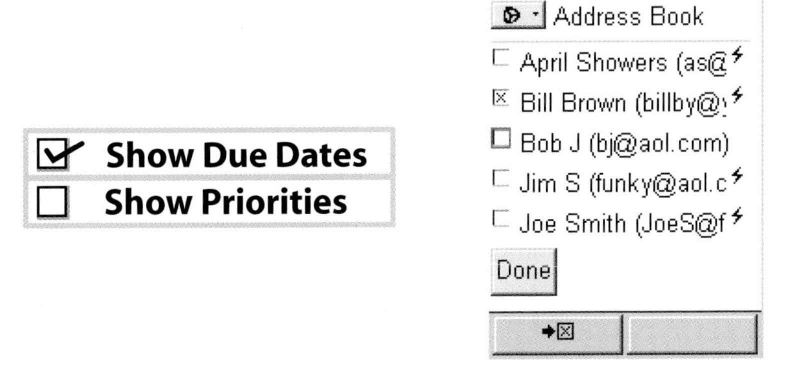

Illustration 4.2 Palm OS® check boxes (left) and Openwave check boxes (right). Palm OS® check boxes reproduced with permission from Palm Inc. Openwave check boxes reproduced with permission from Openwave Systems Inc.

box's state does not affect any of the others. Check boxes appear adjacent to text or graphical labels. Some user interfaces overload checkboxes to show an intermediate state, often as a gray fill rather than a blank or 'X'. Use your judgment – such user interfaces are rarely ideal.

Radio Buttons

Radio buttons always appear in groups, since they are exclusive sets of on–off controls (Illustration 4.3). Only one radio button in a group may be on at one time. A radio button can never appear alone – in that case, a check box would

Priority: 1 2 3 4 5

Sort by: Priority Due Date

Illustration 4.3 Palm OS® radio buttons (left) and Openwave radio buttons (right). Palm OS® radio buttons reproduced with permission from Palm Inc. Openwave radio buttons reproduced with permission from Openwave Systems Inc.

be more appropriate. Palm OS software uses a different appearance for its exclusive on–off controls, which they borrowed from the Xerox Star, the pioneering graphical user interface developed by Xerox in 1981.

Push Buttons

Push buttons look and work the same as they do on the desktop. They are metaphorically similar to hardware push buttons. They initiate an action when clicked. When including push buttons in your user interfaces, follow all of the guidelines from labeled and softkey buttons. In addition, follow the conventions on feedback, naming, placement and state discussed below.

Feedback

Provide audio or visual feedback indicating a button press during the 'press' process. Inverting the colors is a common feedback mechanism.

Naming

Name buttons according to the action they perform, rather than using 'OK' or 'Yes'. Users often quickly scan to find an action button and press it without thinking what that button might be. If you tell them clearly in the button itself they are more likely to succeed. Ellipses ('...') at the end of a button name indicate that clicking the button will result in a dialog box being displayed (see the section on Dialog Boxes, pages 96–97).

Placement

Remember that in most cases users will be tapping the display to activate push buttons. Placement logic changes based on this simple fact, since reaching to press a button at the top of the display will obscure the entire contents. The tradition in most handheld devices is to array push buttons horizontally along the bottom of the display. However, for devices that feature an actionable status bar at the bottom of the display it is sensible to position the action buttons one row up from that bar.

Here are provisional guidelines for button placement on handheld devices:

- Place buttons at the bottom of the display, left justified, when possible.

- Place the most logical action button to the left.
- Place the 'Cancel' button rightmost. The 'Cancel' button closes the current page without accepting any of the changes made by the user. If the changes are not cancelable, the label of the button should be 'Close' instead.

States

Buttons can be active or inactive. Inactive buttons are typically dimmed in some fashion. The standard for black-and-white display of dimmed icons is to mask them with a 50% pixel pattern.

Buttons are made inactive when their functionality is temporarily unavailable. If it is unlikely that your user will understand how to make the button active, an alternative is to keep it active at all times and to display a helpful message if it is clicked in a situation in which it would otherwise be inactive.

Progress Indicators

It is important to provide feedback to users during lengthy operations. Progress indicators (sometimes called **therms**, short for thermometers) are often the best choice for this type of feedback (Illustration 4.4).

Guidelines for progress indicators are as follows:

- Use a progress indicator for any operation that will last more than two seconds.
- Use a single progress indicator for all of the processes in a series. Do *not* use successive progress indicators. A single progress indicator conveys a single delay. Multiple progress indicators are annoying and unhelpful.
- Double progress indicators, where one shows the current process and a

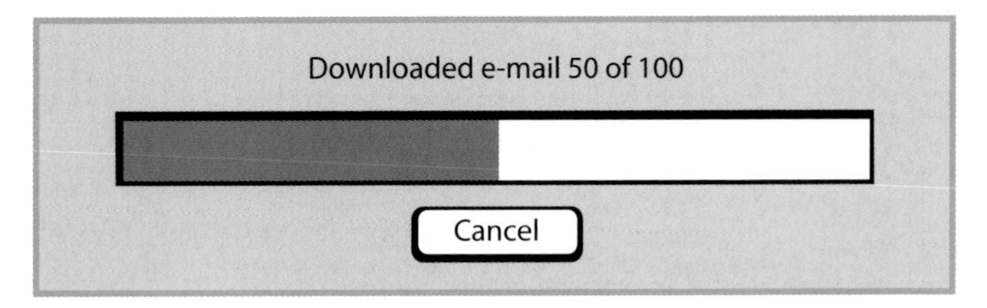

Illustration 4.4 A progress indicator provides feedback during lengthy operations

second shows the progress for the entire series, are a popular 'solution' to the problem posed by the previous guideline. I prefer a single progress indicator with information about the current step. Also, double progress indicators require additional processing power, making the user delay even longer.

- It is helpful to show the current step in a series in text adjacent to the progress indicator. However, if the steps increment so quickly that users cannot read them they are frustrating rather than helpful. If a step takes less than half a second to complete, do not detail it. Just increment the progress indicator.
- If possible, provide a 'Cancel' or 'Stop' option. 'Cancel' would restore the system to the way it was before the operation, while 'Stop' would merely discontinue the process. In the case shown in Illustration 4.4, 'Cancel' would erase the 50 emails already downloaded.
- When possible, if the process will take longer than 10 seconds, indicate the time spent and estimate the time remaining in the process. Update the estimate every 5 seconds in order to keep it accurate.

If you cannot use a progress indicator, such as in the case of a download of unknown size, use an animated graphic instead of a progress indicator. Good options include hourglasses, spinning globes, and spinning clocks.

User Interface Constructs

User interface constructs are combinations of graphical user interface controls, much like a sentence is a combination of words. Following are common user interface constructs found on handheld devices.

Dialog Boxes

A dialog box is a window that appears on top of an existing page, requiring immediate attention (Illustration 4.5). Dialog boxes may be in the form of error messages or information panels. Dialog boxes usually offer 'OK' and 'Cancel' as options and often offer additional options. However, I recommend using specific actions as labels rather than using 'OK', as the action associated with pressing 'OK' can be ambiguous, as in the frequently encountered example of 'Save' and 'Don't Save'. If 'OK' and 'Cancel' were used instead, what would 'OK' mean?

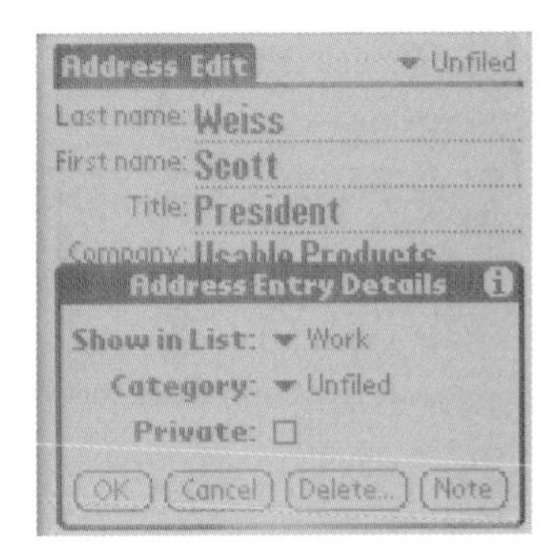

Illustration 4.5 The 'Address Entry Details' dialog box. Reproduced with permission from Palm Inc.

Forms and Wizards

Forms are a collection of labels and text entry fields on the same page. Wizards take those entry fields and place them on successive pages. Wizards tend to be easier to use than forms, but that tendency changes as the number of variables increases. Wizards are also less useful when only some of the variables need to be modified; the wizard user interface forces users to page through each and every variable whether they need to or not.

Clipboard Model

PDAs and pagers support Cut, Copy, and Paste commands, otherwise known as a **clipboard model**. Mobile telephone handsets typically do not support this feature. I feel that a clipboard model is critical to the success of all of these devices, but so far I have had trouble convincing browser developers of my point of view. The counter-argument is that it is difficult to incorporate the clipboard model into a telephone handset.

The clipboard enables users to transfer data easily between applications, although implementation of this user interface on a phone poses challenging problems:

- How would a user select the data to be placed on the clipboard?
- How would the user retrieve the data from the clipboard?

If the device you are developing for does *not* support a clipboard model, consider enabling common transfer actions when users view data records. An example:

> Sally is entering in her calendar an appointment with Joe. She wants to add Joe's contact information from her address book.

On a desktop application, Sally would open her address book, find 'Joe', copy his information, switch back to the calendar, and paste it in. The copy–paste activity might involve several steps if the address application breaks up the data into multiple fields. Each field's content would have to be copied and pasted separately.

On a phone, Sally is not going to be as fortunate, as she will have to triple-tap her way through the appointment process, then save her appointment before switching to the address book, since most phones do not support application switching. Most phones, however, support the 'Talk' button from anywhere, such that if a telephone number is on the display the phone can call it, whether or not the user is in the address book. This direct-action support is terrific, but only gets us part of the way to a superior user interface. Some of the earliest electronic calculators supported memory cells, allowing users to store numbers with the 'M' button (and recall them with the MR – memory recall – button), but phones have abandoned this model. Perhaps it is time to resurrect the 'M' button.

Types of User Interfaces

Fluency with user interface elements provides us with the building blocks, or foundation, of a user interface. In this section, a discussion of common user interfaces will provide you with the basics, while illustrating the challenges and opportunities for designing for handhelds.

Web Clipping

Web clipping strips, scales, or otherwise modifies content that cannot be displayed on a particular device. An alternative to WAP or cHTML, Web clipping enables handheld devices to browse pages designed for the desktop Web. However, many Web clipped pages are nearly useless, especially those with large amounts of related data spread throughout the page. Small displays cannot effectively navigate large pages of data, and Web clipping applications may make matters worse by stripping graphics that are critical to understanding the content.

Home Page

All handheld devices have some type of home page. A home page is a starting point for the user much like the desktop is on a desktop computer. A home page on a desktop website is the 'welcome mat' for visitors. The home page of a handheld device is the place that offers all the major features. Home pages are implemented in different ways but are usually a set of labeled icons. Home pages are designed to be easy to find:

- Phone: 'Menu' key or graphical button at the top of the display
- Palm OS: silk-screened home button or key
- Windows CE: 'Start' button
- Motorola Wisdom™ OS: top-level screen, accessed by travelling 'up' or 'back' from a running application
- RIM Wireless Handheld™: press the 'Back' hardware key from any running application; multiple presses if you are browsing the Web

E-commerce

Electronic commerce for the wireless Web, often referred to as **m-commerce**, is presently modeled after e-commerce for the desktop Web. Desktop e-commerce has been phenomenally successful, with consumers spending billions of dollars per year online. However, very little online spending is being done via handheld devices, because of usability and other problems:

- Entering shipping information on a handheld device is tedious.
- It is impossible to represent merchandise effectively on a tiny screen.
- Remember, people are on the go – they are unlikely to browse from their handsets. Purchasing strategies will be highly targeted.
- People who feel compelled to make a purchase while away from their desktops have the choice between purchasing by phone and purchasing by handheld using the wireless Web. Only if the wireless Web is faster and more

convenient than sales representatives (and hold times) will their purchases migrate online.

Ways to reduce barriers to e-commerce on handsets and PDAs are as follows:

- Offer cross-platform registration. Your customer may want simply to re-order items from their previous order, and if they cannot access that order via their handheld you may lose the sale.
- Support simple passwords, such as numeric-only, for handsets.
- Remember logins and passwords from session to session. Users do not share handsets, so security is less of a concern on the wireless Web than it is on the desktop Web. Handset theft is, of course, a problem, but that is a carrier issue and not an individual application issue.
- Be realistic about what can sell on a handset. Transportation tickets are a likely purchase, but sweaters are not. Any item requiring visual inspection will have trouble selling on a handheld device. Even a beautiful color display on a handheld device will only be a couple of inches wide.
- Establish a micropayment strategy with the carriers. Billing through the carrier will be the most effective means of implementing wireless Web e-commerce. In this way, no account information needs to be entered on the handset; information retrieved or products ordered via handset incurs charges that appear on the carrier's regular invoices.

Usability will be the key to successful m-commerce. Convenience, efficiency, and thoughtful design will lead to success.

Games

Entertainment and gaming are ideal applications for handheld devices. The devices are nearly always in hand, and games provide a quick-and-easy entertainment mechanism when users are bored. Games that enable users to start and stop with breaks in-between are even better, since the time they have to spend may be brief. Arcade-style games are also well-suited to handhelds, since the processing power, controls, displays, and sound generation are all present. A next step for handheld device arcade games is to publish high scores via the wireless Web, to bring a sense of community. Multiplayer games will come next.

A caveat: wireless Web games are at present less than inspiring because of the display and connectivity issues of current browsers and second-generation (2G) networks. As sophisticated user interface controls and animation are incorporated into Web browsers, more interesting wireless Web gaming will be possible.

Game design is outside the purview of this book, but I have included books in the Bibliography that are helpful (for example, see Bates and Lomothe, 2001; Rouse and Rubczyk, 2001). Please remember, though, *prototype your game design and test it with your target audience.* Prototyping a game may require a desktop prototyping application, such as Flash, but the true test will be on the handheld itself.

News Readers

News readers are user interfaces that deal primarily with text and still images. The following are guidelines for news readers, oriented toward handsets:

- Keep headlines short, as they are likely to be used as menu items. If necessary, use the first sentence to clarify the headline.
- Indent paragraphs with three spaces rather than using blank lines. Blank lines could be confused with the end of your article.
- Link concepts within an article to other interesting content.
- When using images, embedded audio, or video, keep the size down to a minimum, especially for 2G network delivery.
- Within an article, set the primary softkey to 'More', and the secondary softkey to 'Hdlns' (headlines). You may wish to use 'Main' instead of 'Hdlns'.
- At the end of an article, provide a list of related, linked headlines. Set the primary softkey to 'Next', and the secondary softkey to 'Hdlns' or 'Main'.

For PDAs, you have more options, because of their larger display size.

Video Viewers and Audio Players

Video viewers deal with moving images that are intended as education or entertainment – even for business presentations. Audio players do the obvious:

they play music, news, audio books, and so on. With the advent of peripheral ports and Bluetooth™ wireless networking, new types of media devices are becoming available on a regular basis, but the device manufacturers have not created standard user interfaces.

Most current handset browsers do not support video or audio, but some PDAs do support these media types. For this discussion, let's focus on non-Web media applications. In a media player, provide the basic functions of stop, play, and pause. Offer rewind and fast forward if you can. Best yet, offer a scroll bar control so that your users can position the cursor where they want in the file. Use the standard icons, too. When possible, display the current position in seconds, and the total play time. Obviously, streaming media may not have a 'total play time', and so such a display would be inappropriate.

File management on PDAs is limited and has usability challenges. Refer to the device's user interface guidelines before forging new ground in user interface design. Consider user testing the recommended implementation before innovating on it.

Specialized Business Applications

By 'business applications' I mean custom applications designed to automate a specific business need, such as preparing a bill of lading for a package, inventory control, or monitoring in a healthcare environment.

The guidelines for creating specialized business applications are the same as those for creating any application: define your audience, and develop scenarios and flowcharts, etc. Here are some direct applications:

- When defining the audience, use traits common to the audience. If it is your company's sales force, perhaps ask the human resources director for assistance.
- Model scenarios after existing business logic. Take the goals and methods in use today and adapt them to the handheld medium. You may need to compromise functionality or methods because of the limitations of the technology.
- Prototype the user interface and test its usability with representatives of the target audience.

Broad Audience Business Applications

Rather than using the buzzword, 'b-to-c', I will refer to 'broad audience business applications'. Checking the status of a flight, online banking, and customer service applications require extremely high usability, since getting technical support might be difficult or impossible:

- Provide consistency with desktop Web user interfaces, especially those by the same provider. In other words, if the desktop website has a specific order process, do your best to stick to it. Users will be alienated otherwise.
- Link the systems so that user profiles are available from the desktop Web to the wireless Web. Allow use of the same user name, password, and customer profiles.
- Accommodate starting and stopping a process without losing data. For example, provide returning users the option of continuing a previously interrupted order.
- It may not be necessary to provide all of the information and options in the wireless Web version of the user interface. Start simple, and add features and options as the market requires.

Productivity Applications

The first generation of handhelds included calendars and address functions. Not until recently, however, have spreadsheets become usable on this platform.

- Not all of the features of the desktop version of the user interface will fit on a handheld. Choose carefully, focusing on data access rather than data input.
- Users are likely to use files transferred to handhelds for reference, so make data-seeking functions available at the top level, not hidden in dialogs.
- Enable users to insert content from address books and calendars into your application without needing to copy each field independently, if possible.
- Provide an easy way to jettison data from your application, in order to make space for more content.
- Provide an option to clear data from the handheld but to restore it at the

next synchronization operation. In this way, space can be freed up temporarily.

Communication Applications

Communication applications include email, video telephony, interactive chat, and asynchronous discussion boards and messaging systems. Until recently the sole function of the pager was messaging. In the first iteration of the pager you were simply notified of a message. Then they became one-way, then two-way, and now there is packet switching so that data can always flow without the need to set up a connection.

- Consider storing the content on the server and storing only the headers on the device to save memory and download time.
- A great feature on Eudora, a Palm OS email package, is the ability to download the first portion of each message. When viewing a foreshortened message, Eudora places a link at the bottom enabling continued download at the next synchronization. An even better implementation would offer the ability to grab the rest of the content immediately, either as an alternative or instead.
- Link address books to your communication application's functionality. Users will want to establish a connection or send a message from within the address book, so make the extra effort to accommodate that need.

Advertising

Advertising may be brochures or purchasing guides, such as interactive decision support systems. Advertising also includes banners, integrated content, and advertorial and disruptive messages. Although few are fans of advertising, it is a user interface that is necessary to consider. Advertising on wireless is an interesting challenge, since there is barely room to convey the primary application's meaning and user interface, leaving even less room for secondary messages.

The following list provides some ideas that may help you in your advertising efforts.

- Include items in menus that bring up sponsors' websites. Be careful to prevent your users from feeling 'hijacked', but instead convey in the menu item that they will be taken to a sponsor.

- Place interstitial pages with ads between content pages. This method is particularly effective in news articles; an ad with links could be inserted between article pages. Readers would prefer that ads be placed at the end of articles rather than between pages, but your advertisers may be willing to pay more for more prominent placement.
- Place ad links at the bottom of the page, such as [Sponsor: E-Soup], [E-Soup], or [Got Soup?]. Place bottom-of-page ads *above* links to other content. Placing the ads above the links enables the advertising content to be scanned as users read the page, without making linking to related content awkward.
- SMS (short message service) ads and alerts can be useful as well. However, these ads could quickly become annoying. You do not want to irritate or alienate your audience.

Permission marketing

You can provide opt-in choices on WAP or Web-clipped sites for SMS alerts to mobile phones – with periodic incentives or based on proximity to a retail outlet. Be sure to provide value to your audience with your SMS and alert advertisements and promotions, as you might be diverting their attention from something important. Good options include coupons, special offers, and notifications of temporary discounts.

Synchronization Issues

Synchronization is the act of reconciling data between a handheld device and a desktop computer. Synchronization requires software on both the desktop and the handheld, and a data connection, either tethered or wireless. Data transfer rates for synchronization are much higher than for wireless Internet access and afford large transfers of content to handhelds reliably and quickly.

Most synchronization methods involve a **cradle** or cable connection, but newer devices utilize infrared, IEEE 802.11, or Bluetooth™ RF (radio frequency) connections to achieve wireless connectivity. Either way, the following guidelines should be applied:

- Provide feedback. Let your user know the connection has been made through a visual and/or audio indication.

- Keep the user abreast of the synchronization process. Show an animated or frequently updated message about the progress of the data transfer. Providing a **therm**, or progress indicator, helps users tremendously. Show *x of y records remaining*, or some analogous message adjacent to the progress indicator. Be sure to use only *one* progress indicator, and not a series of progress indicators: one for upload, one for download, etc. Users do not care how cool your progress indicator is; they simply want to get an idea of how much time remains before they can grab their handheld and go.
- Let your user know when the process is complete with a visual and/or audio indication.
- Forgive user error. If the user terminates the connection early, revert to the presynchronization data set if possible. If not, display an error message indicating how the situation can be corrected.
- Offer opportunity to handle conflicts in a batch, rather than individually. Let users choose to 'accept all changes' in addition to allowing them to resolve each issue individually.

Globalization and Localization

Globalization is the act of designing a product that supports multiple languages and cultures. **Localization** is the act of customizing a globalized product for a particular language and/or culture. A globalized user interface follows the following principles:

- White (empty) space exists within the user interface for 'expansion'. German phrases take up more space than French or English phrases, for example. Character-based languages (such as Mandarin) are often more demanding than Western languages.
- Icon images should be replaceable. Different cultures ascribe different meanings to so-called 'standard' symbols. The palm of the hand, for example, is offensive to Muslims. Mailboxes have different appearances in different areas of the world, as do power switches and electrical plugs.
- Bitmaps should rarely or (better) never include text. Editing a bitmap is more trouble than editing a text file.
- Store user messages and field labels in friendly text files or databases, not embedded within software code.

- Use system software calls to display quantities that use units of measure. The system software will make the necessary translations to the local temperature measurement, time display, or numeric display standard.

The best way to be certain about the cultural appropriateness of your user interface is to employ multicultural usability testing. While you should screen for members of each culture in which you plan to sell your product, having quality assurance testers of different national origin will help even more. Bear in mind that globalization saves time in localization, and localization enables your product to reach a greater audience. Products may be unusable for your intended audience if you do not address these issues.

Designing for WAP for Mobile Phones

WAP is a challenging beast to design for. WAP can be compared to the desktop Web for versions 1 and 2 of the initial browsers from Mosaic, Opera, Netscape, and Microsoft. Web pages looked different on every browser, and the browsers were being released every few months. Site designers had to deal with whether or not a browser supported tables, frames, DHTML, Flash, etc. WAP versions have come out frequently, but most WAP users cannot upgrade their browsers – the handsets must be replaced. Furthermore, Openwave, Nokia, Microsoft, Handspring, and other companies make their own browsers, which adhere to different sets of rules. A given manufacturer may even have different browsers on different phone models. Finally, browser versions are also quite varied. The result is that there are no guarantees and there are many variables.

You cannot guarantee screen placement of your controls, the size of a display, or even whether graphics can be displayed. My advice is to approach your design challenge in one of two ways: either design for the least common denominator or target one specific handset model. Regardless of which option you choose, the guidelines presented in this section will apply.

When designing for a phone, there are several important principles that will enable your mobile applications to provide more value to their users:

- Cross link: provide duplicate links in different areas of the application. Do not simply design a 'tree', since getting from branch to branch requires returning to the 'trunk'. Design a mesh, so that getting from one area of the application to another is easy. It is not necessary to duplicate every feature everywhere; just duplicate the features most likely to be desired by your audience in each particular instance.

- Provide content on every card. If every page has content, then the feeling of 'endless hunting' dissipates.
- Provide links at the bottom of every card.
- Avoid text entry whenever possible. Employ selectable choices rather than requiring text entry.
- Have a maximum of 9 menu items, or 8 and a 'More' menu item. Most WAP browsers automatically place digit accelerators (1 to 9) next to menu items, but the tenth item does not get an accelerator.
- Use consistent, natural language.
- Use title case for menus and softkeys. ALL CAPS IS HARD TO READ.
- Make your case in the first 12 characters. More than that will require 'Times Square Scrolling' – when the text is longer than will fit into a single line, and the handset blinks portions of the text in sequence.
- Use 'Main' instead of 'Home'. 'Home' is ambiguous – is it the carrier's portal, or your application's start page?
- Don't use 'Menu' as a softkey label – it is an ambiguous term. Is it the phone's menu, the browser's menu, or that of your application? Instead, use 'Main'.
- Keep menus as flat as possible. The more hierarchy, the less usable.
- Provide a way out on every menu. Remember, your user may have arrived at a given page by mistake. Providing them a way to start over is a way to keep them in your application. Otherwise they will have to step through a lengthy series of 'Back' operations or leave your site entirely.

I have observed respondents in usability tests getting hopelessly lost once they selected a wrong menu item. The 'Back' functionality is not as intuitive on phones as it is for the desktop Web, so providing a 'Go Back' menu item will help your users.

Metaphor Discussion: Deck of Cards

Wireless phone displays suggest the metaphor of a deck of cards. The metaphor is derived from the fact that wireless displays are small, rectangular, and can be flipped back and forth like a card deck. Although this metaphor is helpful for developers, I feel that it is confusing for designers and users alike. Most people think of WAP cards as pages, and WAP decks as sites. I prefer to stick with familiar terminology, although I prefer 'applications' to 'sites'. I think of brochures when I think of 'sites', but I think of interactive functionality when I think of 'applications'.

Card Types

From the user's perspective, there are four card types for WAP browsers: Display, Entry, Choice, and 'no display' (see Illustration 4.6). An additional card type is 'No Display', which is useful only to programmers.

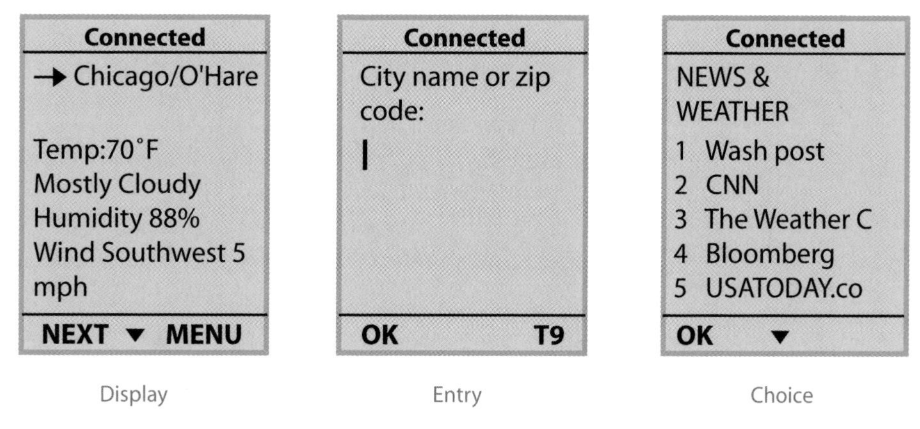

Illustration 4.6 Card types: display (left), entry (middle), and choice (right)

Display

Display cards simply display text, graphics, and links. Sometimes this card type is referred to as 'content'.

Entry

Entry cards display content and entry fields, which are typically indicated by square brackets ('[,]').

Choice

Choice cards enable users to select one of a list of items. The WAP browser automatically numbers the items on these cards.

No Display

These cards are useful mainly to provide the appearance of a short URL, as they do not show up on a user's phone. They are primarily useful for 'redirect', where a simple URL takes the user to the 'No Display' card, which then redirects the user to the final destination. For information architecture design, ignore this card type.

Abbreviations

Since softkey labels are limited to 5 characters, you will frequently need to abbreviate words. Abbreviation is acceptable in WAP applications, although great care should be taken in your creative abbreviation efforts. Following are guidelines for shortening words:

- Use conventional abbreviations when possible: 'Minimum' becomes 'Min'
- Abbreviate from right to left: 'Calculate' becomes 'Calc'
- Abbreviate one syllable at a time: 'Voicemail' becomes 'Vmail'
- Eliminate vowels. 'Tomorrow' becomes 'Tmrw'
- Employ or create an acronym: 'Value Added Tax' becomes 'VAT'

Graphic Design Opportunity

Quite simply, graphic design opportunity is limited or nonexistent on mobile telephone displays. The options you do have include:

- Lower-case characters to create a more casual feel.
- Dash-and-space characters to improve layout – but be careful to not create any blank rows on telephone displays, which could indicate the bottom of the page.

Options you do not have, or cannot guarantee, are:

- font choice
- font size choice
- character attributes, such as *italics*, underlining, and **boldface**

To be completely honest, most users would prefer that you focus on the information presentation and not worry about the visual appearance. The 'beauty' of your application should be its ease of use and its utility to its audience.

Common User Interfaces for WAP

Menus

Menus are lists of choices (e.g. see Illustration 4.7). Phones automatically number these lists of choices, so you should not duplicate their efforts. Most guidelines recommend providing 9 choices and '[More]' for the 10th choice. I *disagree* with this strategy and prefer 8 choices with [More] as the 9th. In this way, users can press 9 if they know what they want is on a following page, while the [More] option requires clicking the scroll button 8 times.

Illustration 4.8 depicts Openwave's new graphical Web browser's menu capability. Openwave has incorporated a menu into its latest browser that works much like a menu bar. The user navigates to the top of the screen with the cursor and activates the menu with an action button click. The menu may then be navigated with arrow controls and a second action button click.

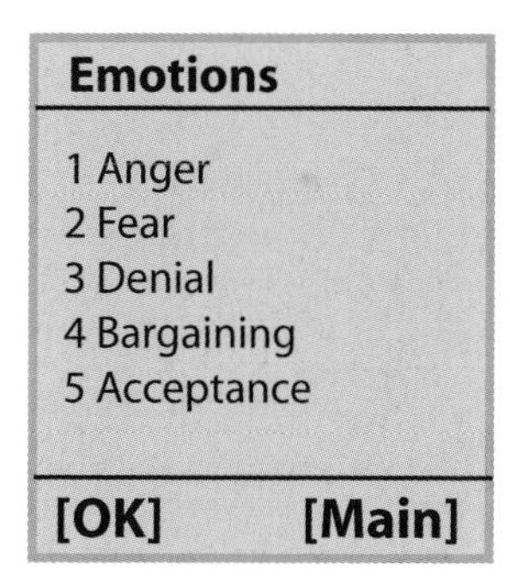

Illustration 4.7 WAP menu

Illustration 4.8 Openwave's graphical Web browser menu capability: start state with up arrow clicked to get to next state (left); menu highlighted, right softkey clicked to get to next state (middle); menu displayed (right). Reproduced with permission from Openwave Systems Inc.

Forms

Forms consist of labels and text entry fields. In Illustration 4.9, selecting (1) the 'Bark' field produces a (2) menu result page.

Dialog Boxes

Preferences and choice pages can be collectively referred to as 'dialog boxes'.

Articles

Access to article-type content from a site should be in the form of headlines rather than a numbered or dated list. If it is breaking news, indicate 'NEW'. If it is weeks old, label as 'archived'. Of course, the article should contain the date

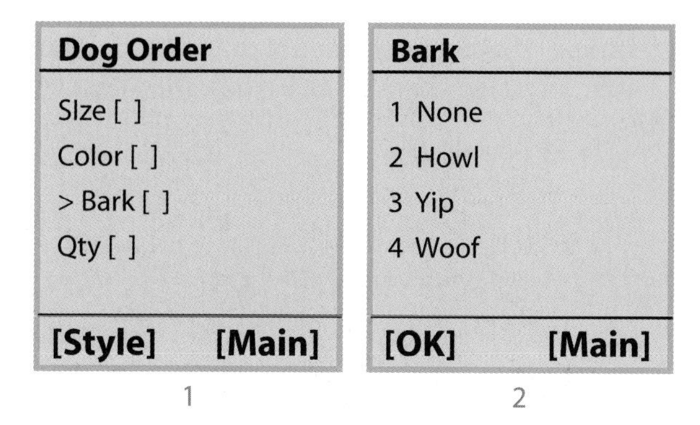

Illustration 4.9 Forms

information at the beginning, but the headline does not need to contain this information.

Articles can be singular or a part of a series. If your site is primarily content, then you may want to present links to additional articles at the end of the selected article. One of the softkeys should take the reader back to the headlines (or the point from which the first article was launched), and the other softkey should take the reader to the next article.

Successive Menus: Alternative to Text Entry

The series of screen shots in Illustration 4.10 shows a process for selecting 'Des Moines' from a successive menu. Successive menus enable users to scroll-and-select, rather than triple-tap, their way through data entry. For cities in the USA, the first choice could have broken the country into regions rather than alphabetical distinctions, but either method is suitable. One site I located had the entire list of US states (50 plus Washington, DC) on successive pages – the worst state user interface I have yet seen in a WAP application. As a rule of

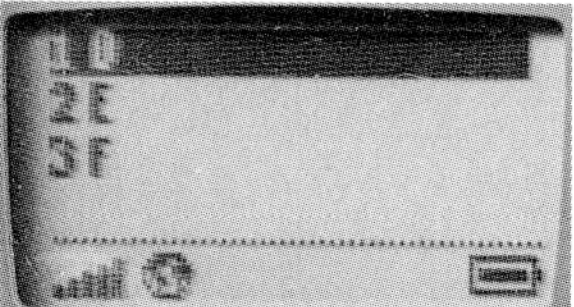

Illustration 4.10 Successive menu approach to selecting a city (source: The Weather Channel)

thumb, if a category is more than 10 items, break it down further. The (good) user interface above comes from The Weather Channel.

Designing for Communicators (PDA Phones)

What Are Communicators?

Communicators are mobile telephone handsets with PDA functionality. They can be modified pagers, modified phones, or devices designed from the ground up to be communicators. Be aware that users will think of communicators as phones first and PDAs later. People use their fingers to dial and terminate calls; requiring users to pull out a stylus or introducing additional steps for these functions is a sure way to alienate users.

Guidelines for Communicator Design

Good communicators require tight integration between the address book and the telephony capabilities of the device:

- Enable actions on address book entries. A user who looks up an entry may want to call that person. Additionally, they may want to send an SMS message or an email. Provide access to all three options.
- Enable one-handed access to all telephone functions. Most mobile telephone handset users use only one hand to dial a call.
- Make telephone actions finger-enabled. Pulling out a stylus to place a call is annoying.
- Make the terminate call function a hardware button if possible. Making call termination a two-step process is extremely frustrating to users.
- Enable 'call' functionality whenever a properly formatted number is highlighted, whether it be in the address book, an email, the calendar, or otherwise. Remember, the device is primarily a telephone. Make telephony primary.

Designing for PDAs

PDAs behave more like desktop applications than like phones since their displays are larger and with higher resolution and they offer richer data input mechanisms. However, PDAs are still small and portable, and PDAs have constrained text-entry capabilities.

PDAs support layered windows, sound effects, and bitmap graphics. They also support multiple applications and a browsable environment from which applications can be launched, similar to the Windows™ desktop or the Macintosh™ Finder.

File System Considerations

A file system enables users to manage data without having to start by launching an application. The 'file manager' is the application dedicated to user data management. File managers allow users to place files into folders, which can themselves be placed in folders. File managers allow the renaming, deleting, and copying of files. Without a file system, each application has to manage its own data – a structure that makes interapplication data sharing difficult.

A stark difference between the Windows CE and Palm OS platforms is in the way they address data management. Windows CE has a full file system, while the Palm OS environment lacks one. However, when designing applications for Windows CE, refrain from generating a large number of files. Not only is memory limited, but the small display size makes it difficult to manage a large number of files effectively.

One of the most significant functionality compromises in the Palm OS environment is its lack of a central file system. Each application must manage its own data separately, and users can access data only from within the application that manages it. Third-party file management solutions are available, such as FilePoint. However, although they provide users with access to files, Palm OS applications were designed without that capability in mind. A file system needs to be integrated into the operating system in order for applications to benefit from it effectively – it cannot be tacked on.

Menu and Command Bar Design

Commands that need to be present most of the time are placed in push buttons. Menus provide access to commands that can be hidden most of the time. Menu commands are typically verbs that initiate action. The Palm OS menu is accessed by tapping the title or by tapping the menu button. Pocket PC menus are present on the screen at all times, indicated by a down-pointing triangle, much like popup menus. Cascading menus, a staple of desktop applications, are a poor idea on handheld devices because of the lack of screen real estate. I do not recommend them. Cascading menus are hierarchical menus, meaning that at least one of the menu items produces a child menu. These menus are difficult enough to use on the desktop; they would be almost unusable on a handheld device.

Command Bars and Toolbars

In the Palm OS environment, the title area of the display has room at the right where push buttons can be placed. These buttons are analogous to desktop application toolbar buttons and are useful for modifying selections. Alternatively, a popup menu can be placed in this location. The Address Book application has a popup menu that switches between the different categories, such as 'Business' and 'Personal'.

Windows CE supports combining the menu with tools and buttons, calling the combined element a **command bar**. Menu items appear at the left, and tools appear at the right, as shown in Illustration 4.11.

The Palm OS user interface supports a command bar, which is a toolbar that appears at the bottom of the display area for 4 seconds. The command bar appears when the user strokes the command accelerator, which is a diagonal line from the lower left to upper right of the Graffiti® text area.

Illustration 4.11 Windows CE command bar example – but it is 483 pixels wide. Reproduced with permission from Microsoft

Tap-and-hold

Both Palm OS and Windows CE environments support 'tap-and-hold' to display a popup menu. The user action is obvious – the stylus is tapped and held in position over an object. After a brief delay, a popup menu appears. If the user wishes to cancel the menu, he or she simply clicks elsewhere on the display. Tap-and-hold menus should reflect the object that they are affecting and should always include the clipboard commands: 'Cut', 'Copy', 'Paste', and 'Clear'.

Dialog Boxes

Dialog boxes provide secondary controls that are best left off menus and push button bars. Place infrequently used controls on dialog boxes in order to reduce screen clutter.

A graphical button with trailing ellipses ('...') launches a dialog box. A dialog box has the following elements:

- recognizable border
- title, at the top
- content
- action buttons, left-justified at the bottom

Floating Windows

I do not recommend implementing floating windows, otherwise known as **palettes**, in handheld applications. There simply is not enough screen real estate. If you must provide secondary controls in buttons, use panels that appear like popup menus that are accessed from a button at the bottom of the screen. The panel should contain all the functionality needed and will be easily displayed or hidden by way of the popup menu control.

Adapting Desktop Applications

Never port a desktop application directly to the handheld platform. 'Porting' an application is translating the software code to a new operating system, while making only minimal changes. Rewrite the application *in the spirit of the original* desktop application. Desktop applications rely on multiple windows, the clipboard model, and a lot of screen real estate. While PDAs support the clipboard model, the 'copy' command is never as easy as the 'control-C' key sequence that so many of us use on the desktop.

Reduce the feature set of the application you plan to port to a handheld device. Screen real estate, memory, and user attention spans are all less on handheld devices than they are for desktop applications.

In the two screen shots shown in Illustration 4.12, Microsoft Excel was pared down for the Palm OS and Windows CE platforms. Sheet To Go was developed by DataViz®. Pocket Excel is from Microsoft.

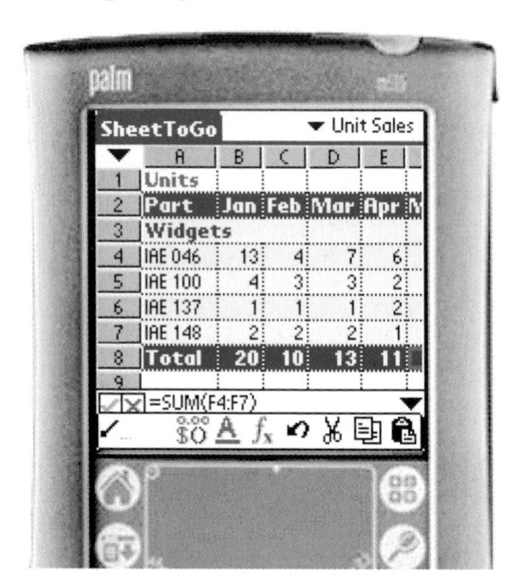

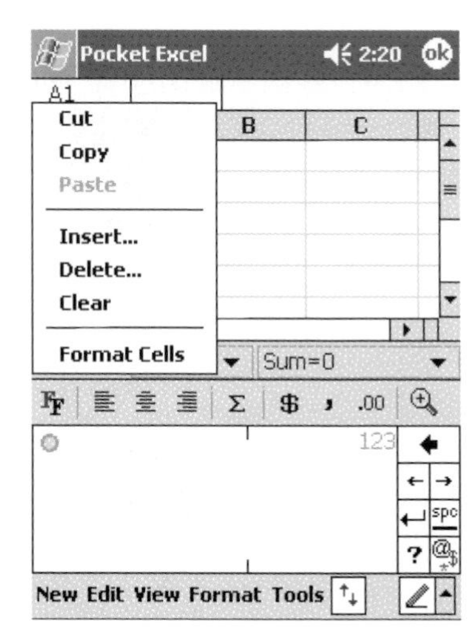

Illustration 4.12 DataViz® Sheet To Go application for Palm OS® (left); Microsoft Pocket PC Excel (right). Image of Sheet To Go reproduced with permission from DataViz Inc. Image of Pocket PC Excel reproduced with permission from Microsoft

Web Design

Web clipping takes standard desktop websites and strips them of whatever the destination browser cannot appropriately render. The screen shots shown in Illustration 4.13 are of Eudora's Web-clipping of eBay and Yahoo.

The biggest complaint I hear about Web clipping is that the first page or two of every clipped site tends to be navigation. You can see that in Illustration 4.13, and also in the example in Illustration 4.14, the example in which is not stripped of content but is instead reformatted for the small screen. Blazer™ does not strip large graphics; instead it shrinks them to fit. This strategy works well, I think, for Yahoo but poorly for MapQuest, which has an auxiliary website designed for small displays (available from www.wireless.mapquest.com/Palm/v3.0/index.html). Maps that are easy to read on a large display are unreadable when shrunk to 160 × 160 pixels. There are times when panning a large image is preferable to shrinking it, but unfortunately Blazer™ does not offer its users the opportunity to choose.

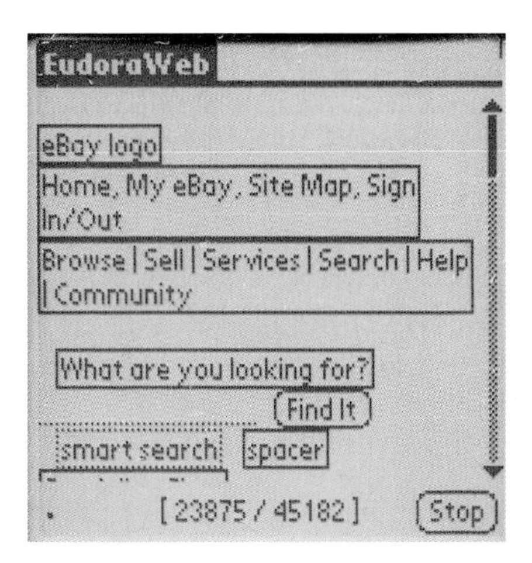

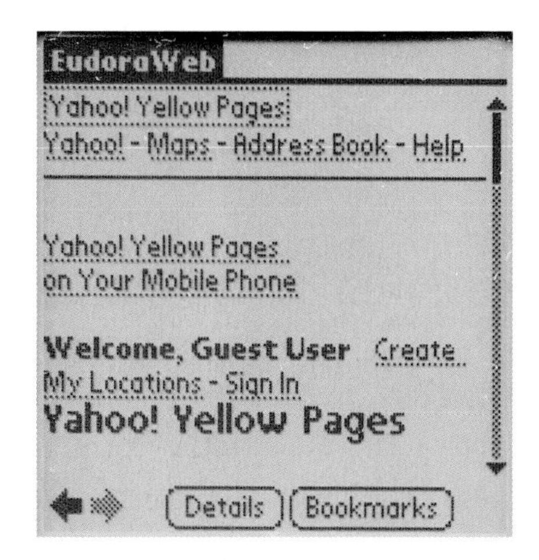

Illustration 4.13 Eudora Web-clipped eBay.com (left) and Yahoo.com (right). Eudora image reproduced with permission from Qualcomm. Image of Web-clipped eBay home page reproduced with permission of eBay Inc. Image of Web-clipped Yahoo! home page reproduced with permission from Yahoo Inc.

Illustration 4.14 Yahoo! as presented by the Handspring Blazer™ Web browser. Handspring image reproduced with permission of Handspring. Image of Yahoo! home page reproduced with permission of Yahoo Inc.

Designing for Pagers

Pagers are two-way email communication devices with QWERTY keypads and that lack touch screens. Two examples are shown in Illustration 4.15. Beepers are not addressed by this book. Beepers are the historical antecedent to

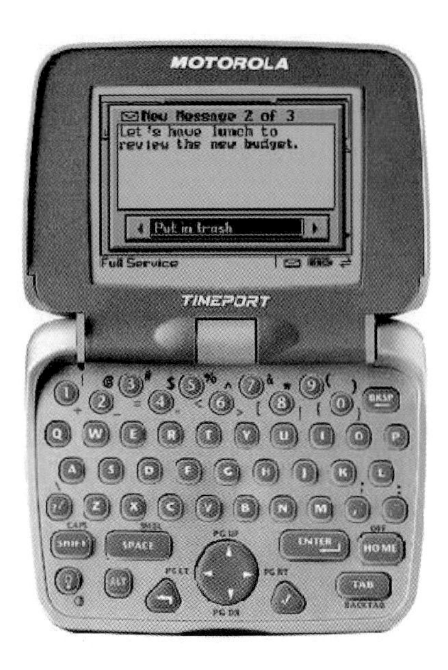

Illustration 4.15 RIM Wireless Handheld™ home screen (left); Motorola's Timeport® P935 mail display. Image of RIM Wireless Handheld™ reproduced with permission of Research in Motion Ltd. Image of Motorola's Timeport® P935 reproduced with permission from Motorola Inc.

today's pagers. Beepers had one-line displays that provided limited information, such as simply the telephone number of the caller. Today's pagers are sophisticated two-way communication devices that utilize email protocol. Some pagers support Web browsing as well, and some pagers have made the leap to become communicators, such as the BlackBerry 5810 Wireless Handheld, which is a mobile telephone handset as well as a two-way email pager.

Pagers offer more user interface controls than do phones, but fewer such controls than PDAs. The fundamental difference between a pager and PDA is the availability of a touch screen. Pagers do not have touch screens, so there will not be a 'pager PDA'.

The biggest challenge for pager design is the lack of a touch screen, which limits functionality to keyboard navigation. The user interface model is primarily to identify a field and enter or modify its options. Designing for these devices may seem simple, but it is in fact more challenging, since navigation, actions, and options are all selected using either a navigational device such as a roller wheel or directional arrows.

The 'Focus': Navigation without a Touch Screen

A pager's user interface is fairly sophisticated, but it doesn't have direct manipulation because no touch screen is available. The RIM Wireless Handheld™ has a roller wheel and 'Back' button that enable navigation. The Motorola Timeport® has a directional keypad, 'Home' key, and 'Back' key for navigation. Both devices rely on a user interface **focus.** The focus is the on-screen indication of which control is active.

RIM OS Design Considerations

The RIM OS user interface has a fairly simple interaction mechanism: the roller wheel (see Illustration 4.16). All interactions are in vertically oriented lists, although a particular list item can have more than one horizontal component. Each page of a RIM OS application has a title bar and a content area. The content area contains fields at the left with their associated values to the right.

Illustration 4.16 RIM Wireless Handheld™ Model 957. Reproduced with permission from Research in Motion Ltd.

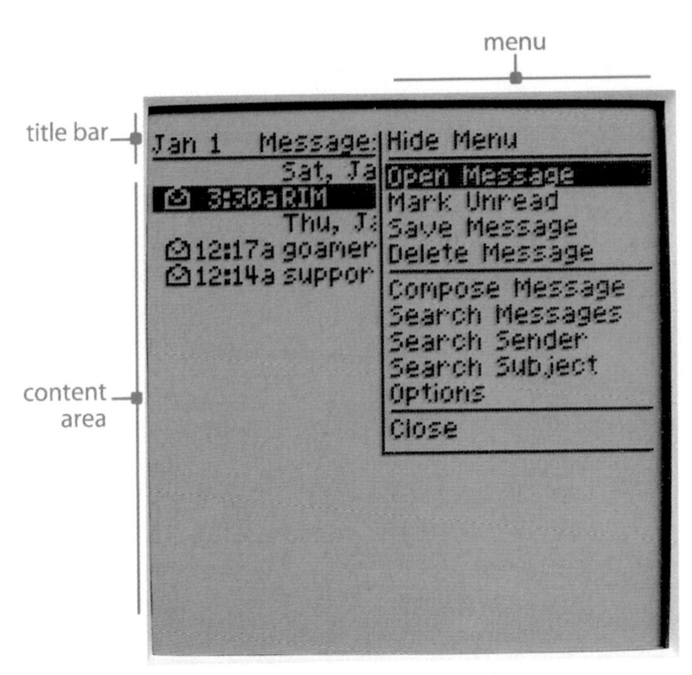

Illustration 4.17 RIM OS user interface architecture. Reproduced with permission from Research in Motion Ltd.

The roller wheel enables the user to perform the following operations:

- roll wheel: moves the cursor
- click wheel: displays the menu
- orange 'Alt' button + roll wheel: changes highlighted option in forms

The interface architecture is shown in Illustration 4.17.

Menus

The menu is context-sensitive, based on the highlighted field. The menu appears at the upper right of the display as shown in Illustration 4.18. Menus are broken into sections with horizontal rules. Menus are *not* hierarchical. The first menu item is always 'Hide Menu', which is a 'cancel' operation. It merely hides the menu. The last menu item is either 'Close' or 'Cancel'. 'Close' is displayed from a top-level application field and when clicked it closes the

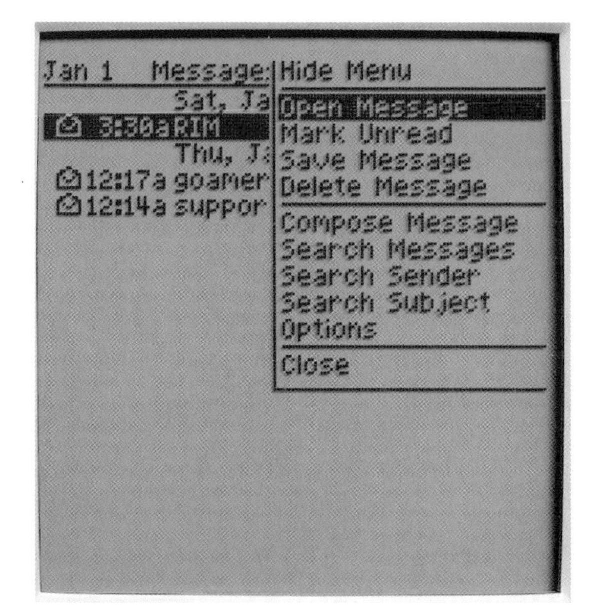

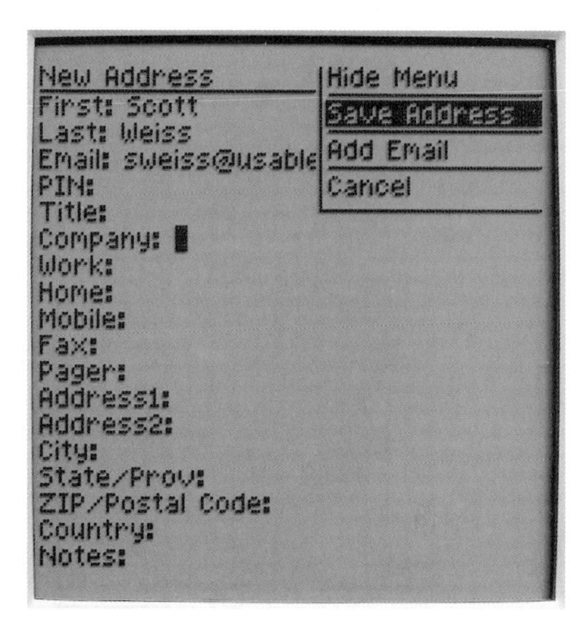

Illustration 4.18 RIM OS application menu (left) and secondary page menu (right). Reproduced with permission from Research in Motion Ltd.

active application and returns the user to the home page. 'Cancel' is displayed from a second-level page, which restores the settings to what they were prior to launching the secondary page and backs the user to the main application.

Forms and Data Entry

Most RIM OS applications use **forms**, which are application pages that consist of labels and values (Illustration 4.19). Labels are placed on the left, values on the right. Text-entry fields have left-justified values, whereas other fields have right-justified values. The roller wheel moves the focus. For text-entry fields, a block text-entry cursor appears, indicating that the user can type in data.

For nontext-entry fields, clicking the 'Enter' key displays a dialog that enables the user to modify the associated value. The orange 'Alt' key acts as an accelerator for this function. Clicking and holding the orange 'Alt' key while moving the roller wheel skips the dialog step and results in the value being changed while the roller wheel is moved.

The action of pressing the orange 'Alt' key while moving the roller wheel is not intuitive, which is addressed by instructional text in most forms, as shown in Illustration 4.20.

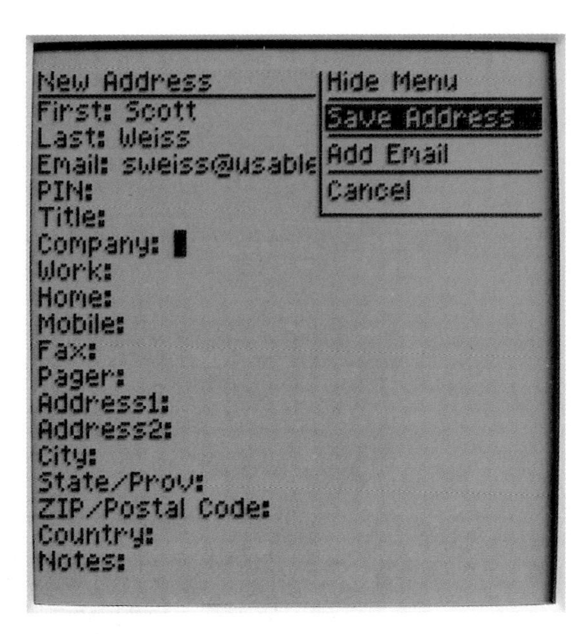

Illustration 4.19 RIM OS: the block text entry cursor is adjacent to the 'Company' field. In this illustration, the menu is displayed at the upper right-hand side. Reproduced with permission from Research in Motion Ltd.

Dialogs

A RIM OS dialog is a rectangle that floats over the underlying content on the display (Illustration 4.21). Dialogs have a prompt and a value. The value is changed by moving the roller wheel. Clicking the roller wheel dismisses the dialog. Clicking the 'Back' button cancels the operation.

Clipboard functions

RIM OS provides access to the clipboard through an nonintuitive orange 'Alt' key sequence. While in a text field, the user is required to click the orange 'Alt' key and the roller wheel. The device then enters 'select mode', during which moving the roller wheel highlights text. A second click on the roller wheel displays the menu, which offers 'Copy' as an option.

Additional information

See www.developers.rim.net/knowledge/pdfs/motient/ui.pdf, specifically the

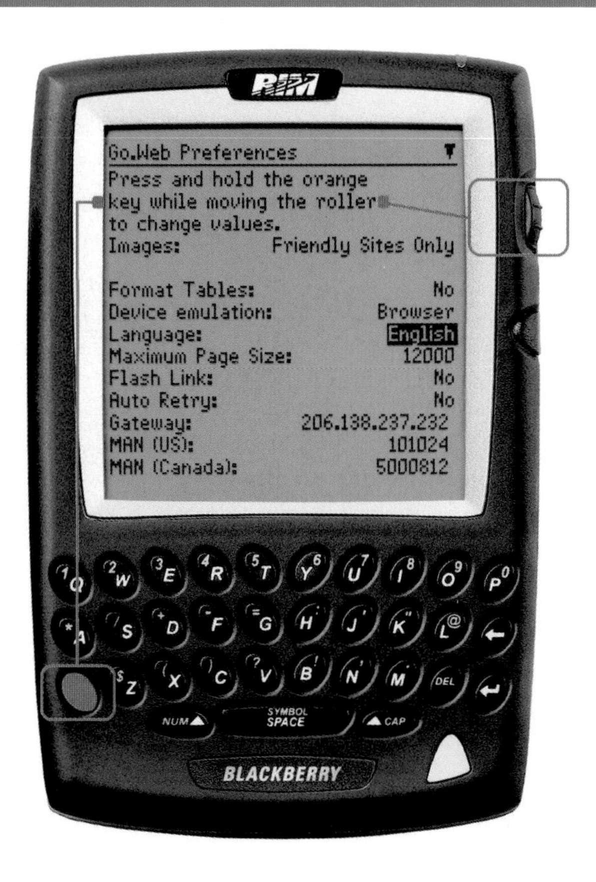

Illustration 4.20 RIM OS properties screen. Reproduced with permission from Research in Motion Ltd.

UI Engine Overview, for further information on the RIM Wireless Handheld™ user interface.

Motorola Wisdom™ OS Design Considerations

Motorola Wisdom™ OS is a rich user interface supporting graphical controls, such as radio buttons and check boxes. The appearance of user interface elements in Wisdom™ OS is somewhat different from that of other operating environments, giving Wisdom™ OS applications a unique look and feel but without sacrificing usability. The Wisdom™ OS user interface has been tuned to the keyboard, since the Timeport® P935 (Illustration 4.22), like all pagers, lacks a touch screen.

Illustration 4.21 RIM OS dialog. Reproduced with permission from Research in Motion Ltd.

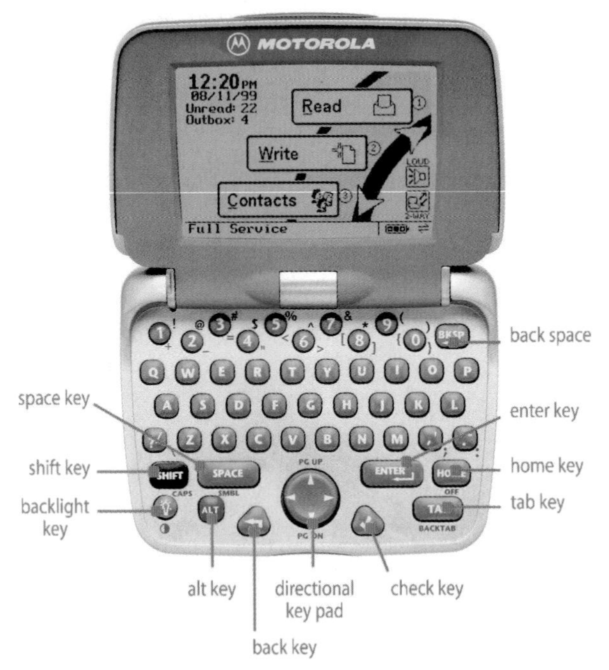

Illustration 4.22 Motorola Timeport® P935 keyboard functions. Reproduced with permission from Motorola Inc.

User interface architecture

Each 'page' of a Wisdom™ OS application has a title bar, content area, icon function list, and a status bar (Illustration 4.23). The icon function list contains the functions that would otherwise be hidden in a menu. Navigation is with the directional keypad and the 'Tab' key. Moving the directional keypad up or down moves the focus. When a control is active the 'Check' key, 'Enter' key, and left–right directional keypad presses act on that control. Up–down directional keypad presses move the focus. The 'Tab' key navigates between fields and between functional areas of the page, such as between the content area and the icon function list.

Illustration 4.23 Motorola Wisdom™ OS email list. Reproduced with permission from Motorola Inc.

Title bar

The title bar, at the top of the application, runs the full width of the display.

Icon function list

The Wisdom™ OS SDK describes this area as a 'toolbar', but its appearance and the way it is used is so different from toolbars found in other operating environments that I feel 'icon function list' is a better name. The icon function list contains up to 9 buttons, arranged horizontally. The left-most button is the key action button, such as 'Send' for an email or 'Save' for an address book entry. The right-most button is the 'Back' button.

Status bar

The status bar should always be present as the lowest item on the display. It displays help information, such as a description of the control that has the focus. Most users will rely on the status bar text to identify the function of the icons in the icon function list. System status icons appear right-most in the status bar, and the application should never conflict with them.

Content area

Within the content area, there can be lists of items, check boxes, radio buttons, or entry fields. For complex applications, Wisdom™ OS also supports tabbed content areas. Tabbed content areas are valuable for multipage applications where each page has related content. Tabs are displayed immediately below the title bar. The directional keypad moves the focus between the content area and the tabs. Alternatively, the user can use the 'Tab' key to navigate through the fields in the content area, then through the icon function list, and then through the tabs.

Once the focus is in the tabs, the user can navigate between them with left–right directional keypad presses. The active tab is indicated with a highlight and an indicator arrow.

Tabbed interface

A tabbed interface is a row of icons below the title bar (Illustration 4.24). Each icon corresponds to its own content area.

Radio buttons

Wisdom™ OS radio buttons are unique in appearance, mimicking sliding hardware switches. Once the control has focus, left–right directional keypad presses switch it between 'On' and 'Off' states (Illustration 4.25).

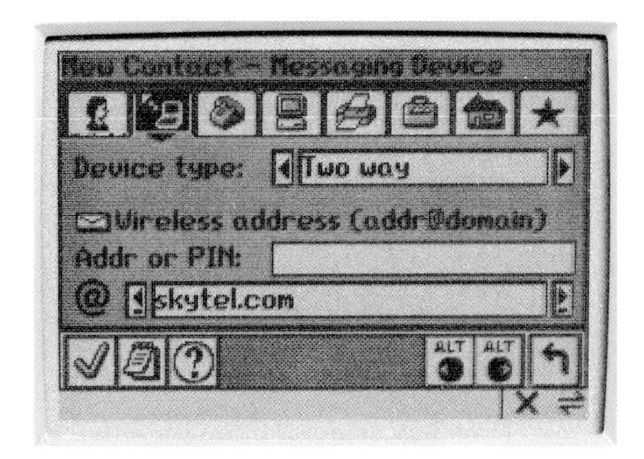

Illustration 4.24 Motorola Wisdom™ OS tabbed dialog. Reproduced with permission from Motorola Inc.

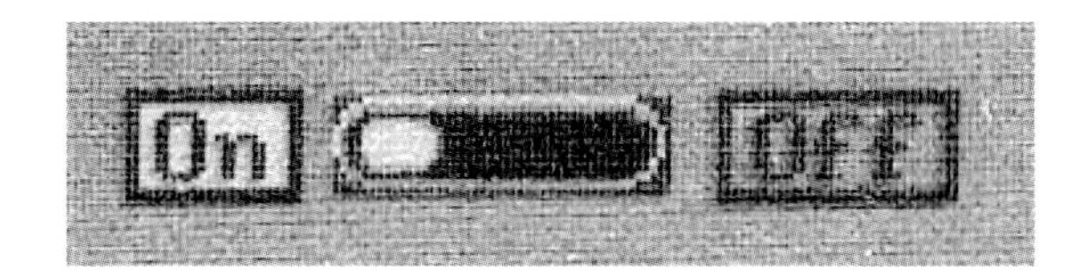

Illustration 4.25 Motorola Wisdom™ OS radio button. Reproduced with permission from Motorola Inc.

Check boxes

Check boxes in Wisdom™ OS appear and work much like they do in most other operating environments (see Illustration 4.26).

Lists

Wisdom™ OS offers designers the ability to check or uncheck list items – an unusual and handy feature. The 'Check' key places or removes a check mark.

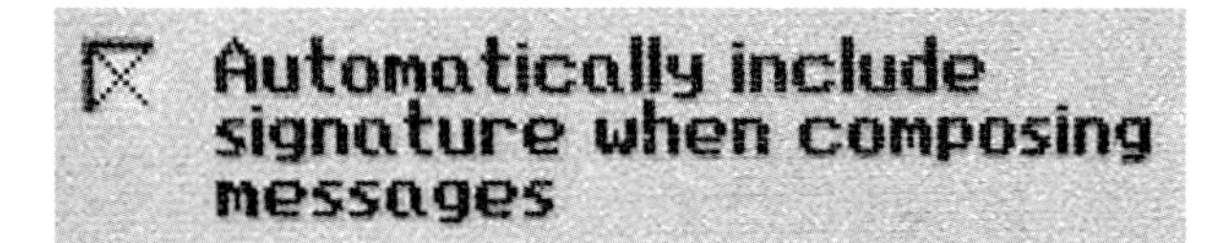

Illustration 4.26 Motorola Wisdom™ OS check box. Reproduced with permission from Motorola Inc.

Pressing ‘Enter’ on a list item with a right-pointing triangle results in a dialog (Illustration 4.27). I was unable to locate information in the Wisdom™ OS documentation describing this behavior, but it seems like a reasonable guideline to follow.

Illustration 4.27 Motorola Wisdom™ OS list with check boxes. Reproduced with permission from Motorola Inc.

Spin list

The ‘spin list’ is The Wisdom™ OS answer to a popup menu. The user changes the value by clicking the right or left arrow of the directional keypad (Illustration 4.28). In Wisdom™ OS, the equivalent of the popup menu is the spin list. In the spin list, the default menu option appears and the user can switch to other menu options by using the directional keypad.

The user can easily determine the number of items in a popup menu by clicking it. In a spin list, in contrast, the user must cycle through each item and remember when he or she reaches the first item again. It would be wise to limit the number of entries to 6 or fewer. By the way, ‘6’ is not a magic number based on research, but a common sense suggestion.

Clipboard support

In Wisdom™ OS, clipboard support is provided by keyboard accelerators (‘Alt’ + C, V, P, or Z), but it is up to the application designer to provide

Illustration 4.28 Motorola Wisdom™ OS dialog with spin list. Reproduced with permission from Motorola Inc.

on-screen control. Include 'Cut', 'Copy', and 'Paste' icons in your icon function list when you can.

Additional information

See the *Timeport P935 SDK On-line Documentation* for additional information, available from www.developers.motorola.com/developers/wireless/downloads/p935_download.html. There is an excellent section titled User Interface Guidelines.

Conclusions

Information architecture practice is similar for desktop computers and handheld devices. However, the constraints of handheld devices make designing applications for them more challenging. Each of the three platforms addressed in this chapter – mobile telephone handsets, PDAs, and pagers – has its own set of user interface controls and how they should be used. Interaction methods differ from device to device, and so each design needs to be rethought for each implementation. Designs for handheld devices should never be directly ported from the desktop but instead should be rethought from the handheld user's perspective.

Summary

Graphical user interface controls for handheld devices are adapted from those on desktop computers. Design guidelines for producing good icons, audicons, menus, popup menus, text-entry fields, check boxes, radio buttons, push buttons, and progress indicators were presented. User interface constructs and combinations of graphical user interface controls were described. Dialog boxes, forms and wizards, and the clipboard model were covered. Types of user interfaces, or applications, were also presented, such as web clipping, home pages, e-commerce, games, news readers, video viewers and audio players, specialized business applications, broad audience business applications, productivity applications, communication applications, and advertising were presented. Synchronization, the act of coordinating data between desktop computers and handhelds, was presented as a user interface and how it can be effectively used. Globalization is designing an application so that it can be easily adapted to different languages and cultures. Localization is the act of customization for each language or culture. Guidelines were presented to make both easier.

The chapter then provided specific advice for designing for each type of handheld device: WAP for mobile phones, and design for communicators, PDAs, and pagers. WAP's most significant limitation is the 12-digit keypad found on mobile telephones, whereas pagers lack touch screens, instead including QWERTY keypads. PDAs have more screen real estate and touch screens but frequently lack QWERTY keypads. Each platform's strengths and limitations require different approaches to solving user interface problems.

Next Steps

Once you've completed your documentation, the next step is to build a prototype of your information architecture. The prototype is built for two purposes: (1) to complete the information architecture and (2) to test usability. Although you might think that you have covered everything in the design phase, building the prototype will require an additional level of detail.

Prototyping

Fear of Prototyping

Well, it is not quite 'fear'. Most software companies acknowledge the value of prototyping, but state that they simply 'do not have the time'. However, had they *made the time* to prototype, they might not have had to delay their schedules to redesign badly conceived products. This logic is common sense, but you would be surprised how many times I have made these arguments to clients.

One builds a prototype of a software application for much the same reason that an architect builds a scale model of a building. The prototype or model gives others the opportunity to experience the final product without spending the time or resources necessary to build the full version. The prototype gives the team, management, and usability respondents a realistic experience from which you can draw conclusions about workflow and usability.

A Prototype is a Proof of Concept

Constructing a prototype enables you to create the illusion of a completed application with a fraction of the time and energy. Team members can evaluate the resulting prototype but, more importantly, the prototype can be tested for usability. Making changes to prototypes is faster and easier than changing shipping applications. There is less resistance among team members to changing prototypes, since in many cases the prototype will be discarded prior to working on the application itself.

Most software projects start with development, continue with the chaos phase, and end up late. The best-run software projects start with a good understanding of the market and the target audience, continue with research and design, continue with prototyping and usability testing, followed by development, and end on time.

A prototype is a working model of your final product. Handheld application prototypes come in two varieties – paper and online. Paper prototypes are made of paper, clear plastic (write-on transparencies) and cardboard; these models require human facilitation to function as 'working' products. Online prototypes are software applications that run on computer desktops or, in latter stages, on handheld devices. Online prototypes function without the need for direct human facilitation, but many of the prototype responses are predetermined (i.e. 'canned').

I feel that all product designs would benefit more from paper prototyping than from online prototyping, at least in the early stages. However, most prototypes are of the online variety. Paper prototypes are faster and easier to build and iterate upon and never 'crash'; online prototypes look and feel more realistic and can be evolved into the final product. The beauty of a paper prototype is that only the best design survives; the danger of an online prototype is that the first version is likely to become the completed product despite severe flaws, as it is too time-consuming and costly to make significant design changes. Consider these facts as you read the following on paper and online prototyping. Note: I rarely recommend producing both types of prototype for a project, as either type is effective.

Getting Started with Paper and Online Prototypes

A common mistake in prototyping is to start with a list of functionality. While it seems reasonable to prototype the *application*, it makes a lot more sense to prototype just what will be tested for usability. Certainly, the application *will* be tested for usability, but step back a moment and consider the application from the users' perspective. What do they want to do? What are the steps that they need to take in order to achieve their goals? These two questions were answered by your scenarios (see the Scenario Development section in the chapter on IA Process, pages 80–81). Take your scenarios and list the functionality required by them, then start building the base user interface elements. Sketch out the data required to accommodate the scenarios, and then place those data, in storyboard fashion, into your prototype. At this point you are just about done, and you need to walk through the prototype step by step to determine what you have left out. List the remaining steps, fix the prototype, and you are ready to go.

Common Concerns

Keep in mind that your prototype models how your product will behave; interactive prototypes are not meant to look and work exactly like finished products. Regardless of your prototyping method, there will be compromises and sacrifices. The two main issues, input and data sets, are introduced here.

Input

Since user data entry (keypad or Graffiti® input), has consistent, poor usability, you may choose not to prototype it at all. However, if you have designed a novel data-entry mechanism, you must prototype it in a hardware or software setting in which it can be effectively tested for usability. Additionally, you may need to test data entry with the keypad or Graffiti® input for comparison, either quantitative or qualitative.

In situations that do not involve innovative data-entry mechanisms, balance the need to test the usability of the whole application with what you can reasonably accomplish in your design or development schedule. The 'whole application', of course, includes data entry, display, and connectivity. Testing the 'whole application' feels better for the client, but paper prototypes are almost always the best choice while the product is still in development.

Data Sets

For your scenarios, you will need sample data. Do not overlook this important aspect of your prototypes, since your clients and your respondents will notice data irregularities and inconsistencies. Sample data take time to generate, as do supporting text and graphic images. Sample data do *not* need to be realistic, but if they do not match a real situation, be sure to make it obvious that you have entered the world of fantasy. Have a little fun with the data, but keep the comedy to a minimum. For Weather Checker, you can make up city names and provide relatively moderate weather conditions – if you find it easier to do so than to look up cities and accurate information. For a financial application, it is nearly always better to find an unused ticker than to use a real company – respondents really care about financial data. Respondents pay closer attention to the data presented on a computer display or handheld device than on a paper prototype, so when using paper you know you will have additional leeway. On the other hand, they will still look at the tickers and complain if it is raining and cold in Honolulu.

Paper Prototypes

The only reason to create a paper prototype – as opposed to an online prototype – is to conduct usability tests on it. If you do not plan to conduct usability tests, skip forward to the section on Online Prototypes (pages 147–151), since paper prototyping will be of little value to you.

The first step in creating a paper prototype is to select scenarios that will enable you to test the features of your user interface. Prior to building your paper prototype, it will be useful to produce a discussion guide, as described in the next chapter. (see the discussion guide section in the chapter on Usability Testing, pages 168–173). Since you do not need to model the entire user experience, you will model only the areas of the interface required to complete the tasks in the discussion guide. For example, if you are prototyping the base weather information in the Weather Checker application, you do not need to prototype the bookmarking functionality. You would include the links to the bookmarking feature but would not build out the screens for bookmark confirmation, deletion, etc.

The idea behind paper prototyping is to test the usability of your information architecture, including that of the hardware device, when appropriate. You cannot test the intricacies of look-and-feel of the shape, tactile feel, and audio feedback of hardware buttons with paper prototyping. In order to test the hardware, you will need to model the hardware and create electronic capabilities that match the functional requirements. However, the hardware models built to test feel and feedback need not support the entire user interface, which will be effectively tested with the paper prototype.

Paper prototyping, being a 'quick-and-dirty' solution, requires compromise. In order to produce a prototype in a few days of a system that will require months to build you will need to leave things out, such as most types of horizontal scrolling.

Each paper prototype will consist of a model of the hardware device and pages from the user interface. I refer to the hardware device model as the 'blinder', since it contains a cutout, behind which the pages will be placed. The pages can be as long as you need, since the blinder limits their visibility. Horizontal scrolling is tough to model with a paper prototype, but it can also be accommodated by increasing the size of the page upon which the device is drawn.

With paper prototypes, the pages that are displayed as a result of hyperlinks or menu selections often contain 'greeked' text. When the actual text for a response is not known at prototype time, designers frequently scribble where the text would appear. As moderator, you can interpret this 'greeking' in an appropriate way for the respondent, or, better yet, ask the respondent what the text might be. The respondent's expectations could drive the design (ergo the term 'user-centered design'). If respondents are confident of what the text will be, their assumptions could easily become the basis for the text in your application.

Dynamic Page Generation

During usability testing of paper prototypes, each page is hand-assembled by team members who act as the 'computer' after each button press or screen tap, since the myriad number of possible combinations usually cannot be prepared in advance. While the pages are being assembled, it is the moderator's responsibility to keep the respondents occupied and/or entertained. The respondents should *not* focus their attention on the act of rendering the pages, but should instead keep focused on the application, perhaps describing what they expect to appear.

The speed of rendering or 'painting' pages will be comparable to the speed of first and second generation wireless networks, but significantly slower than third generation networks. However, the slowness of producing reactions to respondent actions gives the moderator the opportunity to ask probing questions. For desktop paper prototyping usability tests, I always make a joke about the 'speed of the computer', comparing it to a dial-up modem connection, and this joke has proven useful for handheld device paper prototype usability testing as well. Be careful about jokes, though, since offending a respondent is likely to result in the termination of your interview.

At the end of user testing, the prototype has little value by itself. The real value is the act of creating it and testing it, especially since it will change so much between usability interviews. Documenting the paper prototype with a stop-motion video is the final step before putting the paper prototype to rest.

Next I will go through the items you need to have handy in order to produce a paper prototype.

Supplies

The following is a list of supplies that you will need for paper prototyping:

- large format folders (to store prototypes)
- black pens
- transparency markers (permanent)
- transparency markers (wet erase)
- restickable glue stick
- permanent glue stick
- transparent tape
- Post-it® correction tape in three sizes
- white card (heavy cover stock weight)
- write-on transparencies
- colored markers (for color user interfaces)
- manila folders (to organize user interfaces)
- cotton swabs (to erase wet-erase markers)

The magic component in paper prototyping is the restickable glue stick, which enables you to assemble user interfaces quickly and also to disassemble them without tearing them apart.

Components

The first step in creating a paper prototype is to create the display. Take a piece of card stock and draw a rectangle with the same horizontal and vertical proportions as your hardware device's display. Make the size 50%–100%

larger than the actual display, since hand-written characters are always significantly larger than computer-generated characters.

The key component of a prototype of a handheld device is the **blinder**, which is a sheet of card with a drawing of the hardware device with a cutout where the display would be. The size of the cutout is important, since it models the amount of data that can be displayed without requiring scrolling. In order to support scrolling, the card must be larger than the drawing of the hardware device.

Blinders

The 'phone blinder' (Illustration 5.1) is a tool to test your designs. Cut out the box with the 'X' in it, and you've got a simple prototype of a mobile phone. In order to test your cards, mark characters in the grids and place the grids underneath the phone blinder. One way to accommodate scrolling is to use a large sheet of card stock, where most of the card is blank, so that the card can easily be 'scrolled' through the blinder. Another method is to use a sheet of card stock as a platform or 'platter' upon which to place your pages. The full-size sheet of card stock will be reused for each page during the course of the usability interview.

Illustration 5.1 Blinder

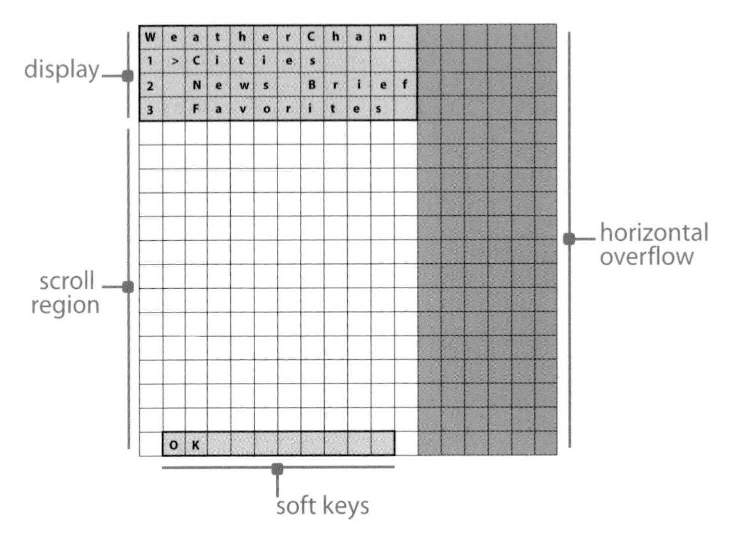

Illustration 5.2 Paper prototype page grid

Phone pages

Illustration 5.2 shows a grid that can be reproduced to use as a form for creating cards. The bold outline represents the visible portion of the display. Each box represents a single character. Any characters outside the emboldened box will require scrolling.

Write the page description on the back of every page, and the date that the page was created. Chances are that you will replace most of the pages in the design, and dating each element will enable you to piece together your design logic after you complete the prototyping process. Apply restickable glue to each page *after* labeling it.

Pulling it all together

Illustration 5.3 shows a series of images demonstrating how the blinder and the pages come together. As you can see, the process is fairly simple. Vertical scrolling is easily accomplished by moving the card up and down inside the blinder. For the cursor, I place a pencil horizontally adjacent to the selected line on the display. Instead of trying to scroll horizontally, I tell the respondents what each line says as they move the insertion point. Do note that horizontal scrolling is primarily used in WAP user interfaces for menus – and horizontally long menus are a sign of poor design anyway.

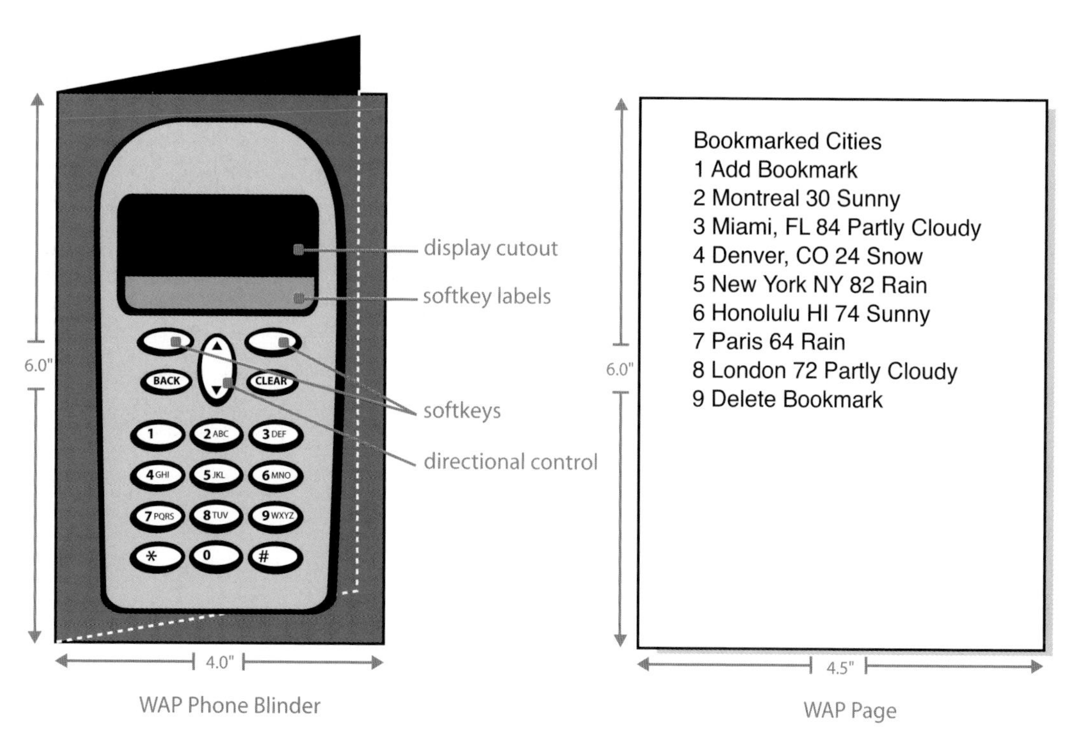

Illustration 5.3 Putting the blinders and pages together: WAP phone blinder (upper left); WAP page (right); WAP phone blinder (lower left)

Step-by-step Instructions

1. Count the number of characters across (columns) and rows on your device's display. For this set of instructions, I use 16 columns by 4 rows.
2. Create gridded cards for the display area. If scrolling is needed, the cards will obviously be longer. I use a word processor's table feature for this step.
3. Measure the 'display' from the gridded card. Take another piece of card stock and cut off a strip about an inch wide from the right-hand side of the card. Draw the 'display', horizontally centered, about one-third down the page. Cut out the rectangle.
4. Draw the rest of the device's buttons, controls, indicators, and silk-screened labels.
5. Using your application map, draw each of the user interface pages onto gridded card stock. Label each card on the back with its description and the date. Draw the softkey labels onto strips, apply restickable glue to the back, and dog-ear the upper right corners, so they are easy to handle. Affix the softkey labels onto a second piece of card stock, and label that second sheet with the title of its associated card. You can imagine what losing a softkey label can feel like during a usability test.
6. Store sections of the user interface in manilla folders.

Paper Prototyping Specific User Interface Elements

Menus: dropdown and popup

Draw the menu control on the page (Illustration 5.4). Most menus have either a shadowed border or a down-pointing arrow or triangle. Be absolutely consistent in your representation of this control, since it will obviously look a little different when hand-drawn compared with computer-generated pixels.

Draw the menu contents on a piece of card with the menu (and the arrow) as its top-most portion. The reason for drawing the menu title again is so that when you place the menu on top of the menu title, you will not have difficulty with alignment. Make sure to fold up one of the corners so that you can easily pick up the menu when you need to. Apply restickable glue to the back of the menu contents *after* labeling it.

Illustration 5.4 A Palm OS® dialog (top left) that I drew for the paper prototyping appendix (Appendix B); the popup menu for 'Work' (right); a filled-in check box (bottom left) adjacent to the 'Private' label

Push buttons

Push buttons are simply drawn on the pages inside of boxes.

Hyperlinks

Hyperlinks are drawn on the pages in the same fashion as they will be rendered on the display – either underlined or in [square brackets].

Text entry fields

Draw a box where the text will be entered and glue a piece of write-on transparency over the box. Cut the transparency piece about twice as large as the drawn box to accommodate messy erasures. You can always draw additional detail on top of the transparency with the permanent marker later.

Lists

Good interface design suggests that lists should not scroll. If your design requires a scrolling list you will feel the pain in prototyping it. Methods to scroll the list include making the page, itself a secondary blinder, and positioning the list contents behind the page, or showing list contents in blocks, replacing one set with another as respondents scroll up and down the page. List selections can easily be identified with an overlay of clear plastic with an outline drawn in permanent marker.

Softkeys

Model softkeys by writing them on small bits of card that get placed with restickable adhesive on the blinder in the appropriate place.

Hardware buttons and indicator lights

Model hardware buttons and toggles by drawing them directly on the blinder. Model indicator lights by creating colored card bits that can easily be adhered or removed from the blinder. Of course, the act of adhering them or removing them is an undeniable clue to respondents – if only indicator lights were that obvious in real life!

Paper Prototypes during Usability Interviews

I always interview two respondents at a time for paper prototype usability tests. I find that the two respondents feel comfortable in the casual environment that paper prototyping creates, but that a single respondent easily becomes overwhelmed by the experience. In the section on the Number of Respondents Per Interview in the chapter on Usability Testing (see page 158), I discuss advantages and disadvantages of one compared with two respondents in usability interviews.

One team member will act as the moderator, asking respondents to complete tasks and querying them about their thoughts during the interviews. One or two team members will act as the 'computer', painting pages and placing and removing user interface elements according to respondent actions.

Respondents often want to join in the fun of operating the prototype, and it is the moderator's responsibility to limit these activities in order to maintain the realism of the user experience. Respondent suggestions for prototype improvements (as opposed to user interface improvements) should always be met with 'Thank you for the great idea. I will discuss it with the team following this interview'. Do *not* argue with respondents, no matter how silly their idea may seem.

Text entry

We already know that text entry is problematic on handheld devices, since there are only 12 keys available to produce 26 letters of the alphabet. I recommend employing 'voice response' text and numeric data entry, having the 'computer' enter text using the wet-erase marker into text entry areas.

Cursor movement

Model cursor movement without requiring individual directional commands. Require respondents to 'point' using their index finger. Unless the device you are modeling supports stylus input, activating a target on the screen will require a softkey press. If stylus input is supported, allow respondents to 'click' simply by saying 'click'.

Hardware scrolling controls, such as roller wheels and up–down directional buttons, can easily be modeled by allowing the respondents to use the extender tabs on cards that require scrolling.

User interface elements

Model each user interface element in its before, during, and after states. Draw the 'before' state directly on the page. Draw the during state (or states) on bits of card, and remember to label each component. Include surrounding detail if necessary in order to make the component practical to pick up and place by hand. Model the 'after' state on strips of card or on resultant pages, depending on the nature of the user interface. For menus, model the title on the page, the menu contents on a card, and each menu item on a strip that can be placed on the page once one is selected.

Page changes (links)

I model a 'wait cursor' for page transitions. I draw an hourglass on an index card, which, when in the mood to entertain respondents, I 'animate' for them by rotating it. Amazingly, nearly every one of them chuckles – the first time I do it.

After All of the Interviews Have Been Conducted

Once you have conducted all of the interviews for this iteration, you *must* document the user interface. An essential first step is to produce a *prototype video*, a stop-motion recording of the paper prototype, narrated with the usability tasks. Video recordings are not high resolution enough to capture all of the detail of most paper prototypes, so the voice-over narration is critical to understanding the prototype video. The process of creating the video, much like the process of creating the paper prototype, is more important than the final result. Going through the tasks one last time, and recording the visuals as you do so will imprint the user interface into your memory. Watching the video a few times while writing the functional specification will be very helpful.

1. Create a voice-over script from your discussion guide. This script will be the spoken narrative accompanying the video. In your script, include a brief project description, the date, and a sentence or two about what the videotape includes. Introduce each task, and identify the pauses, when the user interface will require user actions and/or display changes.
2. Set a video camera on a tripod on a table, on which the prototype has been fixed. This time, tape the prototype down – it is very frustrating to watch a video where a subject that is supposed to be stationary moves around. Train the camera on the prototype. Place a card with the project description and date on top of the prototype.
3. Record the voice-over introduction.
4. Step through the tasks, using the camera's pause feature while staging the prototype after user actions. Use your finger or a pointer to indicate key presses. I tend to use a chopstick, which makes a great pointer.
5. Once you have completed all of the tasks, you are done. There is no need for a set of concluding remarks.

Most prototype videos that I produce are about 10 minutes in length, although the videos take about four hours to produce. The videos are useful to the design and development team, especially for the production of a functional specification. Once the specification is completed, the video is unlikely to be needed further.

Now that you have a great prototype video, write your functional specification. Document each user interface and each of its components. Use lots of visuals.

Next is the discussion of online prototypes. You already know that I value paper prototypes more, but there are circumstances when online prototypes are the right choice.

Online Prototypes

Whereas paper prototypes are quicker to produce, cost less, and are easier to modify, online prototypes are more effective for clients who are more hands-off, or who are very detail-oriented. Online prototypes look terrific and make excellent presentation tools for investors and for focus groups.

One might think that 'comps', static mockup drawings that represent final applications, might be sufficient for prototypes. While comps are great for functional specifications, they are not very useful for usability testing. As a reminder, usability tests measure respondents' performance as they attempt to complete tasks. There are not many performance opportunities with a static image.

If a respondent is faced with a comp of Weather Checker, one could ask the respondent what each menu item does, but one could not ask the respondent to perform a task. In the comp shown in Illustration 5.5, the respondent would most likely say that he or she would click '2' to enter the city, but that does not suggest that the respondent would succeed at completing the task. A counter argument could be that there would be comps for successive pages but, in that

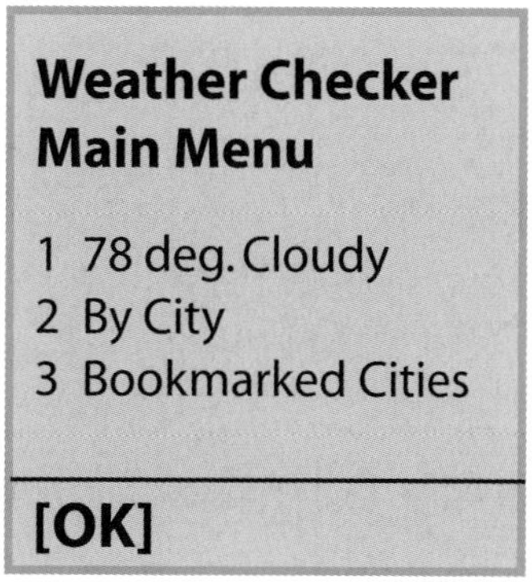

Weather Checker
Main Menu

1 78 deg. Cloudy
2 By City
3 Bookmarked Cities

[OK]

Illustration 5.5 Weather Checker card

case, why not just complete the interactive portion of the prototype and test the user interface effectively?

Clickable demos and functional online prototypes – either on a desktop computer system or on a handheld device – provide a more realistic environment for usability testing than do paper prototypes, although they take more time to produce and iterate upon. Once comps have been turned into clickable demos, a skillful moderator can present them as if they are fully working prototypes. In a clickable demo situation, the moderator must work to create the illusion of true interactivity, much like an actor creates the illusion of a realistic setting on a movie set. An online prototype is a lot like a movie set, as it looks and feels like the real thing for the audience, but only the features that need to function are built, and the rest are simply 'painted on' to provide context.

How to Deploy Online Prototypes

Whenever possible, demonstrate prototypes directly on the handheld device upon which the application will be deployed. When it is impossible to deploy your prototype on the handheld device, consider deploying it on a different handheld device that can simulate your target platform. As a last resort, use one of the desktop emulators. Desktop emulators are even further from the user experience of a handheld device than are paper prototypes, since the desktop environment overwhelms the emulated device.

Clickable demos

Regardless of the type of prototype presented in usability tests, I tell respondents that what they are experiencing is in fact a prototype and that it may not work as well as or like the final product. Respondents forget this disclaimer almost immediately, but stating it initially not only makes clients feel better but it enables me to remind respondents of that fact each time the prototype crashes, performs slowly, or otherwise does not do what the respondent expects.

Desktop handheld device emulators

Emulators are available for WAP (Illustration 5.6), Palm OS® (Illustration 5.7), RIM Blackberry, Windows CE, and Symbian OS platforms. Some emula-

Illustration 5.6 Openwave's WAP emulator. Reproduced with permission from Openwave Systems Inc.

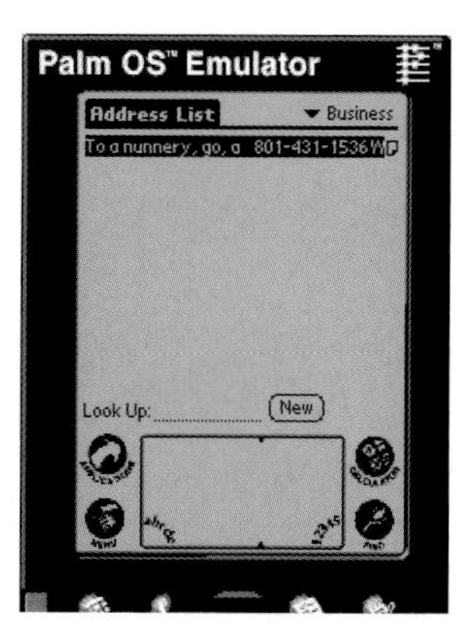

Illustration 5.7 Palm OS® emulator. Reproduced with permission from Palm Inc.

tors are available directly from manufacturers of hardware devices, and others are commercially available for sale. These desktop software products enable developers to test their code without transferring it to handheld devices. The emulators are often fraught with problems, however, since there are always inconsistencies between the emulators and the devices they pretend to be, both in appearance and behavior.

If the device that you need to test on is not available, your only option may be an emulator. Openwave offers a WAP emulator that enables you to test

Table 5.1 Emulators: advantages and disadvantages

Advantages	Disadvantages
Readily available	Looks and feels like a computer display
Quicker to develop and deploy	Input methods are unrealistic
Can be easily videorecorded	Appearance different from actual device

your URLs on several popular phones. However, emulators lack the true feel of a handheld device. Ultimately, the emulators look slicker than paper prototypes, but the interaction mechanisms are actually a bit worse: they misrepresent the devices, making them appear easier to use. They are great for demos, however.

Table 5.1 lists the advantages and disadvantages of emulators, from a usability testing perspective.

Emulators help developers test their code; they do not help information architects try out their designs. For that type of prototype, you will need to use a different type of product, such as creating a website or using a hypertext product such as Hypercard (Macintosh), Toolbook (Windows), or Macromedia Flash.

Creating an Online Prototype

In a hypertext or multimedia product, you can place the screenshot and superimpose invisible hyperlinks or buttons, which can link to other pages within the prototype. In short order, you have created a clickable demo. Your next step is to add conditions and text entry to your linking, which will make the difference between a clickable demo and an interactive prototype.

If your application generates weather information, you can flow in the current date, making the prototype seem more realistic. You can code the up–down buttons to scroll the display as well.

Prototyping connectivity involves measuring the speed at which data flows up and down on the device on which your software will be available. Program in delays so that your prototype matches the speed of your device. You may choose to program in faulty connections to model the 'real world', but be careful to ensure that the 'faults' are realistic and consider placing the faults at the same location for each user test.

None of the emulators or prototyping software effectively recreates the user interface for input, however. All of the desktop-based products rely on the keyboard and mouse as input mechanisms. If you work hard enough, you can locate a portable keypad and code it to emulate a handheld device, but by that time you would have been able to download your application to the device already. However, it may not be necessary to prototype the input mechanism for the device you need to test – you already know what those challenges are.

Conclusions

Prototyping enables you, your team, management, and your target audience to experience your product before it is built. Prototypes are faster to create and less expensive than final products, since they are not fully functional. Paper prototyping is a participatory method of usability research. It is also more fun to do, especially when working with a team. Of course, the most important reason to create a working prototype of your product is to test it for usability, the subject of the next chapter.

Summary

A prototype is a working model of your final product. Both paper and online prototyping formats have been presented in this chapter, with a special emphasis on the paper prototyping method. Paper prototypes are built with ordinary office supplies but require pages to be dynamically created based on user actions during usability tests. Online prototypes can be either semifunctional code on a handheld device or on a desktop computer with emulation software. Clickable demos are a series of static images that mimic what the final application might look like. A skilled moderator can create the illusion of a working product with a clickable demo. Online prototypes and clickable demos have the advantage over paper prototypes of being usable for demonstrations since they look realistic.

Paper prototypes are better suited to teams that work well together – in the same room. Some argue that paper prototypes are not realistic enough, but respondents adapt to them quickly during usability interviews. Interviewing two respondents concurrently alleviates tension and increases the amount of feedback, since the respondents work together. Paper prototypes are easier to change than online prototypes, and can even be changed in-between usability interviews. After interviews, paper prototyped user interfaces are recorded with stop-motion video since the details of how the user interface works are quickly forgotten.

In the next chapter on Usability Testing I go through the steps necessary to prepare for and conduct user tests as well as how to benefit from the data generated by the interviews.

6

Usability Testing

Why Test Usability?

The first time a user experiences your design – frequently in the form of the final release – he or she forms an opinion about the product. If their experience is unsatisfactory, it is unlikely that they will ever use the product again, much less purchase it outright. Therefore, testing for usability should not be an afterthought. If you wish to ensure success of your design in the marketplace, usability testing is not optional; it is mandatory. Whether testing is done in a formal or informal environment, it is best done *early* and *often*.

Readers' Note

This book, as you are well aware of by now, is written in the first person. However, you will see a lot of 'we' references in this chapter, which was written from the point of view of the team with whom I work. In many cases, my colleagues do the bulk of the work, and so our work is represented collectively. I considered referring to the idea of 'in common practice' or using other third-person style, but our process is not common – it was carefully developed through years of practice.

What is Usability Testing?

Usability testing is a process of interviewing people, according to a carefully structured protocol, as they use your product. We recruit test participants, known as *respondents* in market research terminology, on the basis of a written questionnaire, the **respondent screener**. The screener is designed to filter out people who aren't representative of your product's target audience. Once appropriate respondents have been chosen, a slate of interviews is scheduled. During each interview, the test moderator asks a respondent to complete tasks. The essence of usability testing is watching respondents struggle with your designs – with the intention of improving the design so that the product's audience has less difficulty with the final product. Observing how people interact with your product will enable you to identify ways to improve the design, thereby enhancing the user experience and improving the commercial viability of the product. A caveat: this type of observation is not easy. The human instinct is to help people, not to watch them fail. An effective moderator, therefore, will keep a respondent focused on tasks without aiding him or her.

Qualitative or Quantitative?

Usability testing is a qualitative, not a quantitative, measure. Quantitative study involves determining answers to questions based on statistics, whereas qualitative study involves team discussion following an evaluation of empirical data. Many of our clients initially request quantitative study, until they understand the interplay of qualitative and quantitative studies in the design process. When you do not have much time, which is nearly always the case in the rapidly evolving technology marketplace, qualitative study is preferable, as it can be accomplished quickly, efficiently, and cost-effectively.

For those who have need for quantitative analysis once their product is launched, log files of user activity provide ample data for analysis. Log file analysis provides unbelievable opportunity for usability enhancement by way of its quantitative nature. Most logs are overlooked, since technology innovators move so quickly that they neglect the obvious. The logs reveal not just how many visitors came through your site, but where they went and for how long. Location and time information will tell you what paths visitors follow on their way to finding information, completing transactions, and where they opt out.

However, usability testing up front can prevent opt-out behavior and failed attempts and can increase customer satisfaction through design iteration. Designs can be optimized to improve visitor performance before the product is finalized and launched.

Usability testing is fundamentally different from use of focus groups. Focus groups involve gathering several people in a room and asking them their opinions about a product. Usability testing involves interviewing one or two people at a time while they attempt to use a product according to predefined tasks. Usability testing measures performance, while focus groups gauge opinion. Focus groups are not an effective means of gathering usability data, since the group dynamic eliminates the opportunity to observe each respondent using a product individually. Focus groups are most effective during a product's ideation phase or in-between product releases. Usability testing is effective during design and after release. Usability testing *can* derive some of the same opinion-oriented information that comes from focus groups in the form of open-ended questions asked of respondents, but the group dynamic is what most separates the two disciplines. The knowledge learned from focus groups drives the design process with respect to features and competing product analysis.

The following list is an overview of the usability testing process:

- select the facility
- write a respondent screener
- recruit respondents
- write a discussion guide
- conduct interviews
- use the 'think aloud' protocol
- debrief
- document findings
- edit video highlights
- prioritize issues

Usability Testing Timeline

A timeline is provided in Illustration 6.1.

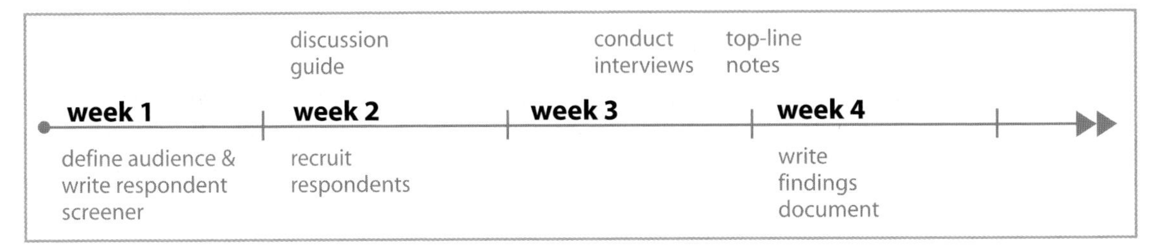

Illustration 6.1 Usability testing timeline

When to Test

Each and every product has a life-cycle. Products go through phases of concept, design, development, and, finally, release. Usability testing has value throughout the product life-cycle, but testing is implemented differently depending on what stage a product is in. Conducting one large study is far less effective than conducting several smaller studies, with design improvements incorporated in-between.

Usability tests are often conducted on released products. However, tests can easily be done on prototypes of products built while designs are being finalized. Testing prototypes is certainly preferable, since prototypes are less expensive to change than released products. Prototypes are quickly put together and far less robust than final products, since they are intended to illustrate a product concept. I recommend testing usability early and often, using paper prototypes initially and graduating to working prototypes later in the process. Testing products that are shipping to customers is an excellent way to determine what to change for the next version, since the test results often indicate that significant change is necessary. If a product were to be tested regularly during its design-and-development phase, fewer problems would be discovered when the shipping product is tested. However, our goal in testing is to identify usability problems, which are always present in products, no matter how well designed. The severity of usability problems correlates to how much testing was done throughout the product life-cycle – and how the results of usability tests influence subsequent redesign.

Justifying Usability Testing: Cost vs. Value

Usability testing helps designers understand areas where people will encounter difficulty, but it has frequently been assumed by management to be time-consuming and expensive. Arguments in favor of usability testing can be found in *Cost-justifying Usability* (Bias and Mayhew, 1994). Usability testing *is* fairly expensive in terms of both time and money. There are ways to control costs, however. One of the easiest cost-saving methods is limiting the number of individual interviews. However, you must strike the right balance between cost and efficacy. If you conduct too few interviews, you will miss important usability problems. If you conduct too many interviews, you will go over budget *and* you may lose sight of the most important problems, becoming overwhelmed by smaller, less significant issues.

'Usability' is currently a buzzword. However, as emerging technologies mature it will become part of the development process, just as quality assurance testing is now. How is its value proven? Quite simply, as companies realize that their products are failing their customers, they will aggressively seek ways to fix those products. Usability testing tells companies how and where their products fail. However, in their eleventh-hour quest to discover how to fix their product, these companies have delayed testing until the most expensive time of all: when the product is already shipping. As was stated earlier, conducting usability research during the design phase can prevent product disaster.

Number of Interviews

Practitioners of usability recommend many different quantities of interviews, for several different reasons. For shipping products, I recommend six interviews, for the following reasons: (1) six one-hour interviews can be conducted in one calendar day; (2) testing six respondents allows you to identify trends. If no one successfully performs a task, a simple 'none' defines the outside parameter. If two out of six are successful, we refer to this as 'some'. If more than half successfully complete a task, we indicate that 'most' respondents succeeded. Remember that usability testing is qualitative. We try not to mislead readers of our usability findings documents. For example, two out of six is more effectively described as 'some', rather than as 33%, since the percentage might imply a greater overall number.

Number of Respondents per Interview

One respondent is suitable for most interviews, especially formal usability interviews conducted in a focus-group facility. Co-discovery, or interviewing two respondents at a time, is valuable in settings where the observers are present in the same room with the moderator and respondent. However, respondents predictably perform much better when in a team, so some usability issues may be missed. Respondents talk to the moderator when interviewed individually, but speak to each other when interviewed together. During co-discovery, they solve problems together rather than seeking help from the moderator. Sometimes when one respondent cannot figure out how to accomplish a task, the other respondent will take over. In such situations, the more capable respondent can dominate the interview. A good moderator will manage two respondents effectively, but interpersonal dynamics are a significant challenge in co-discovery projects. Last, never pair a man and a woman in a two-person interview. Gender dynamics present unnecessary opportunities for trouble.

Types of Usability Testing

Depending on the stage of the product cycle, we offer clients 'accelerated' and 'comprehensive' usability testing alternatives. Comprehensive testing includes transcription and edited video highlights. Accelerated testing lacks those two services. We conduct accelerated tests of paper prototypes in a conference room. We conduct accelerated and comprehensive tests of working prototypes and shipping software in focus-group facilities. The terms 'accelerated' and 'comprehensive' are unique to our process, but you may find them useful in describing your process to your colleagues. There is no industry standard for the services offered in usability testing.

Facility Selection

Facility selection was placed before other steps in the usability testing process because facilities must be booked far in advance. If you do not plan ahead, you may have to wait to conduct your interviews, or you may have to select an alternative facility.

Usability tests can be conducted in a focus-group facility or conference room. While there are dedicated usability labs, at this point in time most of these facilities are designed for desktop software and Web usability tests, and usability lab prices are always higher than for comparable focus-group facilities. Usability lab interview rooms tend to be very small, and the video setups are not always appropriate to testing handheld devices, so the cost to you may be even higher, and the usability lab facility may not have the expertise to help you with your audiovisual needs. If you need a quick solution, a usability lab may be your best option, however. We prefer to use focus-group facilities since we have our own portable video equipment. These facilities tend to be less expensive, more spacious, and are more readily available than usability labs. Market research facilities change constantly, so be sure to research what is available in your area.

Focus-group facilities offer the advantage of separating the respondent from the observers by way of a one-way mirror and acoustic (sound) separation (Illustration 6.2). However, these facilities are rather sterile environments, and they are costly to rent. They enable you to sit with the respondent and have a comfortable interview while your observers watch, safely behind a one-way mirror. You will not be able to hear the observers groan, cheer, type on their laptops, or take telephone calls. They will be able to eat, discuss issues, and come and go without disturbing you. In order for them to see what is going on, however, you will need to arrange for a fairly sophisticated audiovisual setup. (audiovisual issues are covered on pages 173–177).

As was stated earlier, we always test working prototypes and shipping products in a focus-group facility. Having observers present in the room is almost unworkable when interviewing one respondent at a time. Observers always interrupt the moderator, and their facial expressions and laughter are always misinterpreted by respondents who might feel embarrassed and 'shut down' as a result. In contrast, we always use conference rooms to test paper prototypes, but we test two respondents at a time in order to prevent them feeling overwhelmed by the presence of observers. Paper prototype usability testing is much more informal than working prototype testing, owing to the deliberately messy nature of paper prototypes.

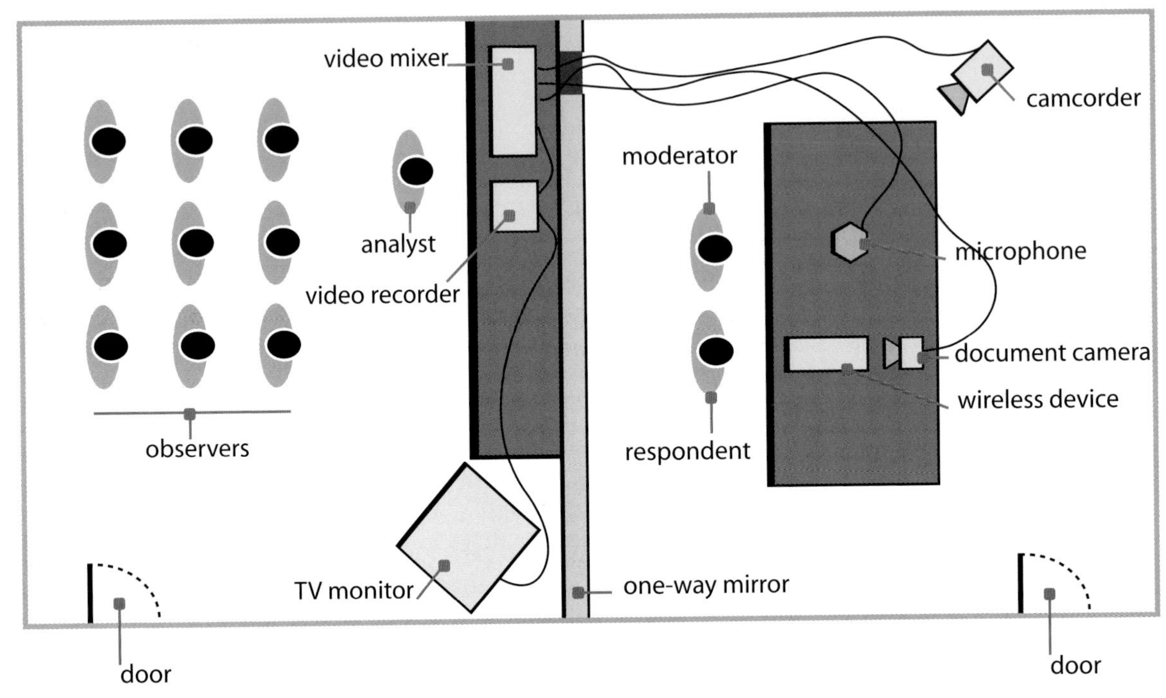

Illustration 6.2 Example of a focus-group facility

Usability Testing of Prototypes: Paper and Online

Testing of online prototypes is best done in a focus-group facility. Testing of paper prototypes is most effectively done in a conference room *with the team members present.* Video technology does not represent paper prototypes effectively for accurate reproduction, and the amount and layout of video equipment required for paper prototype usability tests is excessive. Paper prototype usability testing lends itself to a more casual environment; go with it.

The value in testing prototypes of user interfaces is that the designs can be more quickly and easily modified than can completed application code. Paper prototypes are especially easy to modify, since all of the team members can work on them together in-between usability interviews. Paper prototype usability interviews are typically not done in multiples before changes are made, but instead the design evolves in-between interviews as a direct result of team discussion. This strategy is aggressive and incredibly successful, but not for every team. Online prototypes are more conservative in that multiple tests are conducted on the prototype before design changes and prototype modifications are considered.

Prototype testing always requires a moderator who is able to create the illusion of a 'live' application. Live applications perform calculations on respondent inputs, providing real results. Prototypes usually contain fixed ('canned') responses to respondent inputs. Respondent selection for prototype user testing is the same, but the moderator's ability to 'act' does come into play. A good moderator will be able to describe why the 'inaccurate' results appear, or stage the interview in such a way that the respondent may not even realize that the responses were all predetermined.

You are probably wondering whether or not respondents accept these paper prototypes as 'real' applications. Respondents always have some initial misgivings, but they very quickly adapt to clicking the 'screen' with their fingers, and they especially like the 'voice recognition' text input. The level of comfort and confidence of the moderator and the team directly affect the respondents' comfort and ultimately the success of the process.

When testing paper prototypes, we deliver a functional specification, not a usability report. We sometimes also write a document describing the process that we went through to arrive at the functional specification, but the main purpose of paper prototype usability testing is to evolve the design of the information architecture, not to document the usability problems in the application design.

Write a Respondent Screener

Usability test results are only as good as the selection of respondents. The respondent screener is a questionnaire that will help you or your recruiting partner find representatives of your target audience. When writing this document, keep in mind that you want the *sweet spot* of that audience, not the extremes. Just as the sweet spot of a tennis racket is the place where it is most effective in order to hit the ball, the sweet spot of your audience is someone who is the right age, has the right level of experience using similar products, and has the right amount of intelligence, education, and background. Good respondents fit neatly within predefined ranges of traits, not outside those ranges. Your product will have exceptionally gifted – and intellectually challenged – users, but most of its users will be 'average'. Do not try to accommodate every possible user; focus on those who reflect the widest audience possible.

Take the following example of sweet spot selection for a recipe finder. First, identify the extremes: a university student who has not yet learned to cook – and who might have no interest in cooking – would yield usability data that is irrelevant. On the other hand, the executive chef of a three-star French restaurant would be equally irrelevant. Testing a recipe finder with a homemaker who has a subscription to a cooking enthusiast magazine would likely be exactly on target. The homemaker is the sweet spot.

If your product is for sports nuts, such as a scoreboard, the sweet spot may be a 'sporting enthusiast'. In order to recruit an enthusiast, you will need to screen out people who do not have a working knowledge of the sports represented by your product, and you will also need to screen out the fanatics. Someone who watches soccer every week during a season may qualify, but someone who has not seen a game in two years would not. Someone who plays in a local league may qualify, but a professional player would not. A logical question is, 'Why not?' Someone with no experience with sports will not have the basic understanding of how sports scores are formatted. In order for them to derive value from the product, it will have to provide a lot of background information that may be annoying or frustrating for the enthusiast. The expert will have different interests and is therefore an 'extreme' with respect to this audience.

Audience definition

We place audience definition at the very beginning of the design process (see Illustration 3.1, of the design cycle, in the chapter on information architecture process, page 62). If you defined your audience early on this step may be very quick. If you have not yet gone through a formal audience definition process, this step could raise concerns. However, better to have those concerns raised *before* the usability tests than learning about the disconnect between product and audience from customer support. It is a good idea to review the audience definition every several weeks with your team: marketing is likely to have new ideas; sales has probably revised the requirements based on customer requests; and support certainly has new data from call analysis.

In the audience definition (see the section on Audience Definition in the chapter on information architecture process, pages 72–80), you described traits that represent your target user. In the respondent screener, you need to quantify those traits within acceptable ranges. Most of our clients are initially frustrated with questions such as 'What is the age range of your audience?' However, when we offer a range, they usually want to modify it. No matter what the range is, 25–50, 18–39, 20–40, they want to modify the lower number and the higher number – and each team member usually has a different set of numbers in mind. The important thing is that the age range be defined; the actual low and high numbers are a matter of discussion among team members. Facilitate the discussion and bring the team to consensus, which is more challenging than it sounds.

Nuts and bolts

Start your respondent screener with a paragraph introducing the caller and the project, from the respondent's perspective:

> Hi, I am <insert name> with <insert company>. My client is surveying the public about new features for its next mobile phone. Do you have a few minutes to answer some questions about your mobile phone usage? I am not a telemarketer and will not try to sell you anything.

Until you are certain that the candidate qualifies as a recruit, it is inappropriate to reveal that you are recruiting for a usability interview. If someone does not qualify, just tell him or her that the survey is complete and thank him for his time. Despite the fact that I just suggested that the respondent screener be introduced as a survey to potential recruits, *the respondent screener is not a survey and should not be used as one.* A survey is a series of questions that looks almost identical to a respondent screener. However, surveys are given to hundreds of people at a time and the results are tabulated. For a research project, the respondent screener is used to recruit a small number of people for interviews. Most calls are terminated early and the data discarded. Recruiters will charge significantly more to survey than to recruit, since more time and documentation is required for a survey. If you have any research-relevant questions for your respondents, ask them before, during, or after the interview, not on the respondent screener. The recruiter's job is tough, so try to keep the respondent screener simple.

Ask potential respondents when they last attended a research session, be it a focus group or usability test. There are people who attend research as a supplement to their income – these people are not good research respondents. They might perform well, but they might say inappropriate things during the interview such as 'the last focus group I attended...' Try explaining that to a client. Although it is easy for potential recruits to lie when they are asked about how recently they have attended research, you are better off asking than not.

It is important to screen out people who work in industries where they will know too much about your product or technology. We always screen out market researchers and anyone in the advertising industry, since they tend to know too much about the research process. Many of our clients also require that we screen out journalists who might write about the research. Sometimes we present respondents with nondisclosure agreements, depending on client requirements, but journalists are savvy enough to figure out a way to avoid violating these agreements and still writing about exciting new technology.

Following those two very basic question sets, screening for professional respondents, as well as market researchers and journalists, the respondent screener covers the required traits in an order that is intended to reduce the amount of time spent on the phone by the recruiter. Questions are asked in a ranked order of priority, with quicker, more obvious eliminations made first. For example, if your project requires recruits who are the financial head of their household and whose individual income is between $25,000 and $75,000 per year, first eliminate those who are not heads of household, and then ask about income.

In order to screen out inappropriate respondents as quickly as possible, create questions that have a range of desirability and an unacceptable component. Respondent screener questions can be related and require complicated combinations of traits. Some traits will be critical, and others desirable. Critical traits, of course, need to be screened for first.

Questions are written for the recruiter, who will read them to potential respondents. Recruiters can make up to a hundred calls *per recruited respondent*. The respondent screener must be written in such a way that the call can be completed quickly and easily, but with great accuracy. Answers must be easily filled in or marked with checks or crosses. In the following example, the acceptable annual individual income range is $25,000–$75,000.

Which of the following annual income categories describes you?

I make less than $25,000 per year	Terminate	
I make between $25,001 and $35,000	☐	Recruit a mix
I make between $35,001 and $50,000	☐	Recruit a mix
I make between $50,001 and $75,000	☐	Recruit a mix
I make more than $75,000 per year	Terminate	

Note the 'terminators' and acceptable traits. Also note that the question is not asked as 'How much money do you make per year?' If at all possible, write your questions so that the potential respondent will not be offended. Give them opportunities to say 'yes' and 'no', which are easier answers to provide than actual numbers, which may seem more personal and therefore more of a violation of their privacy. Finally, 'recruit a mix' requires the recruiter to distribute the respondents among the income strata specified. The very best recruiters will do this distribution anyway, but clarifying your requirements up front will ensure a happier working relationship.

If your client or their product is at all controversial, such as a tobacco company, you might consider employing a **favorability rating** question in the respondent screener. It is valuable to screen out individuals who have negative bias about issues related to your interview. Technically, respondents who are opposed to the tobacco industry would be just as useful for usability testing of a cigarette manufacturer's affinity product as those who are ambivalent. However, if a respondent cannot objectively participate in the interview, and he or she uses the time as an opportunity to rail about his or her hatred of the product or company, he or she may not complete all of the required tasks. Worse, your client may insist on the interview being discounted or eliminated from the findings. In order to avoid such unpleasant situations, favorability ratings can be used. Use them wisely, since they are not always necessary. When writing a favorability rating, avoid revealing who your client is or what you are researching. For our earlier example, here is a favorability question:

How do you feel about the following industries?

Intense dislike	Dislike	Indifference	Like	Enthusiastic
1	2	3	4	5
Firearms industry		_____		
Petroleum Industry		_____		
Tobacco industry		_____	Must be 3 or higher	

For this example, you are concerned only about the tobacco industry. However, the other industries are introduced merely to throw off the potential recruit from knowing exactly what the research is about. The answers are given in the form of a numeric rating, called a **Lickert scale**. You can find many Web pages about Likert scales on search engines. I found the definition on Bill Trochim's *Center for Social Research Methods* (www.trochim.human.cornell.edu/kb/scallik.htm) to be helpful.

The respondent screener that was used to recruit for the late-2000 Usable Products Company Sprint PCS & NeoPoint 1000 Usability Study is included in Appendix C.

Recruit Respondents

We often work with professional market research recruiters to perform this task, since recruiting is both time-consuming and difficult. In the event you must recruit respondents first-hand, and don't have the opportunity to hire a professional, here is a set of criteria for selecting research participants.

First, recruiting requires lists of people who might match your respondent screener. Difficult recruits can require 100 or more prospects for each recruit. The more convenient the time and the more appealing the incentive, the easier the recruit will be. However, some types of respondents are just extremely difficult to recruit, especially doctors. In the case of investment professionals, be prepared to deal with skepticism, since, in their line of work, most calls they receive are from solicitors. Remember to be cordial no matter what, since you are representing not just yourself but your company – and the market research industry.

Nondisclosure statements and releases

Respondents may be required to sign release forms and/or nondisclosure agreements. In these cases, be sure to alert them to this fact during the recruiting process, and provide this documentation as far in advance as possible. Requiring a respondent to read and sign an agreement immediately prior to the interview introduces the risk that the respondent will refuse to sign. Some governments may require a signed release informing respondents of their rights, especially with regard to video and audio recording. Check with an attorney if you are unfamiliar with the laws pertaining to privacy in the region where you are testing.

Overrecruiting

Always overrecruit. Research participants are human, and things come up that require cancellation, and, worse, sometimes participants just do not show up. Either recruit two respondents for each one-person time slot, or recruit **floaters** who will cover two consecutive time slots. Be prepared, and have reading material on hand for floaters who might not be interviewed. We recruit nine respondents for six one-on-one interviews. We recruit three respondents for each co-discovery interview.

Incentives

Cash works best and is the simplest incentive. We always use cash. I have seen discussions on usability-focused mailing lists for years about giving away free products, gift certificates, affinity paraphernalia, and platitudes. Let me repeat, cash works best. How much cash depends on the target audience, test times, your region, and even the test location. Try starting with US$50 and go up from there. We have recruited for projects where the incentives were as high as US$250 for a two-hour time slot.

Floaters are paid more than 'regular' respondents since more of their time is required. How much more depends on the project. A good rule of thumb is to pay them 50% more than regular respondents.

All respondents who arrive on time for research are entitled to their incentive. 'On time' is usually defined as within 15 minutes of the scheduled start time of the interview. Respondents who are replaced are almost always paid their incentives, even if they arrive impaired, such as being visibly unwell or inebriated. If a respondent is impaired, take up the issue with your recruiter, not the respondent.

Respondents

Once you have secured recruits for your project, send out letters with directions, a schedule, and a reminder to bring reading glasses if they are needed. Do not just send email, since email is easily lost or disregarded. Send them hard copy via the postal service, which they can bring with them or post on their bulletin boards. Place reminder calls, too, and even place confirmation calls the afternoon before and/or the morning of the research. All of these details may seem overly cautious but, take my word for it, they are all necessary for a successful recruit.

Outsourcing recruiting

If you outsource your recruit, be sure to require daily updates on recruit progress and check the respondent data provided to be sure that they match the requirements of your respondent screener. If not, do not hesitate to require a re-recruit, since without good recruits your interviews could be worthless.

Professional recruiters should take your respondent screener and facility directions and manage the process of selection, updates, and reminders. They usually interface with the focus-group facility to provide a list of names and incentive amounts. Most facilities will prepare incentive checks ahead of time, although they may require a deposit from you. Do not, however, assume that a recruiter will follow all of the steps that are outlined here. Since the recruit is the foundation of the research, if you outsource the recruit you still have the responsibility to manage the recruiter and check his or her work on a daily basis.

Write a Discussion Guide

Once the respondent screener is written, during the recruitment you can begin work on the discussion guide. This document contains background information for test observers, introductory material for respondents, pretest questions, instructions to the moderator for how to set up each task, the tasks respondents will perform, and follow-up questions for respondents. The discussion guide from the late-2000 Sprint PCS/NeoPoint 1000 Usability Study, published by Usable Products, is given in Appendix C.

Background information for test observers

We always include maps to the research facility, since the discussion guides are typically sent out to observers in advance of the interviews. We also describe the **platform** and **test bed**. The platform is the operating system software and hardware on which the product or design will be tested. The test bed is the physical description of the prototype and how it is set up. Test bed information should include descriptions of connectivity enhancements or challenges, mountings, and other test-specific issues.

Introductory material for respondents

Include a series of statements about the interview that will be read to respondents before you start. Here is an example:

> Our client is in the process of designing new features for their product. We want to make sure that the new designs will perform as well as possible. Your opinions and time here today will help ensure that the design is successful.
>
> We are evaluating the product and *not you* in this exercise.
>
> The product you will be experiencing today is a prototype and may bear little resemblance to the final product that will be released to customers. Since it is a prototype, we may experience delays, and at times I may need to reconfigure the device to continue with our interview.
>
> You may pause or end the interview at any time if you are feeling uncomfortable.
>
> Ask any questions that come to mind, but due to the research nature of this interview, we may not be able to answer a question during testing, as it may compromise our results. However, after the study we'll be happy to answer any remaining questions you might have.
>
> Please note that this meeting is being video and audio recorded, which will be used

for internal purposes only.

Please think out loud. Your thoughts are very important to us.

We may be pressed for time, and I may move you from task to task quickly. This is not a reflection on you or on your performance today.

Each of the points made above bears equal weight. Your respondents may not know what to expect, and they are likely to be nervous. The pretest statement lets respondents know what you expect of them and what their rights are. I read the pretest statement verbatim to each and every respondent.

Pretest questions

You've read the pretest statement and you're ready to start the interview. The pretest questions 'warm up' the respondent to the interview. While they take up valuable time, open-ended questions about your respondent's job, experience with technology and specific content, and other project-specific queries can give the moderator and observers valuable information that would otherwise be unavailable. The respondent screening process eliminates people unsuitable for research, but it does not tell you who you have chosen. How do they use their devices? What line of work are they in? Ask them these questions and be sure to include something specific about your project's domains.

Instructions to the moderator

Technical instructions, such as Web addresses, default parameters, login names and passwords, etc. should be included in advance of each task. Place moderator instructions in *italics* to prevent the moderator from reading these instructions aloud.

Tasks

After this warm-up, provide the respondent with his or her first actual task, which should be fairly easy but still relevant to your research goals. Tasks most often come from scenarios that utilize different product features. Most products offer more than one way to accomplish a task, so construct your scenarios with that in mind. Using an example from the scenario development section in the chapter on information architecture process (see pages 80–81), let's develop a task:

> **Using Weather Checker**
>
> *This first task requires the respondent to launch the Weather Checker application.*
>
> You are presently in San Jose. You are going to New York City tomorrow for a business meeting, and are unsure of what to pack. You will use the Weather Checker to find out the forecast for New York for tomorrow.
>
> Using your mobile phone, bring up the *Weather Checker* application.

The discussion guide's tasks must cover all of the functional areas of the product identified for research, which may be a subset of the entire functionality of the application.

The first task does not accomplish much other than requiring the respondent to launch the application. While the task requires a specific application to be launched, the method for locating the application is left to the respondent. The task is undirected, as are most usability tasks.

Usability tasks create an opportunity for the respondent to solve a problem, *not* to follow your instructions exactly. Ask the respondent to meet a goal, not to push a button. I often *do* ask a respondent to do specific things in order to familiarize him or her with the process before introducing more complicated tasks. Write tasks from the user's point of view, not that of the application.

Asking questions

Asking a respondent questions about tasks gives you an idea of their attitudes toward your application. During goal-seeking tasks, ask the respondent questions about the process:

- What do you think about the information on the display?
- What can you do from this point?
- What concerns do you have about this product?

The more open-ended the question, the less you can predict the answer. Sometimes unexpected answers can be very interesting to you and your client. However, respondents occasionally refuse to answer questions that are too vague. Be prepared for anything.

Avoid yes and no answers

Actually, when possible, avoid asking respondents how they would do something and instead ask them to perform a task. Asking a respondent, 'Will the blue button tell you the weather?' gives him a 50% chance to guess correctly. Asking, 'What will the blue button do?' will make him or her think. Another option is to ask 'How can you get the weather from this page?' Instead, go for the imperative: 'Please get weather information from this page'. If the respondent fails, tell him or her to 'Click the blue button and tell me what happens'.

Ask open-ended rather than yes–no questions, as this requires the respondent to think about the application and its process and share more information with you. Do not ask questions such as 'Was it easy to use?' This question will not retrieve useful information. Do ask questions such as 'How does this application compare with others like it that you have used?' This question will require respondents to position your product with respect to others.

Survey tasks

Unlike goal-seeking tasks, which have been covered in this section already, survey tasks require a respondent to answer a series of questions without using the application. You might ask what different words or navigation items mean, or under which category particular functions might be located. We often include a survey task early in the interview to gather feedback about the first page of an application:

- What does this application do?
- How do you feel about this page?
- What other applications is this application like?

Complicated tasks

In my example above, launching the Weather Checker, we only got to the application – we still have not asked the respondent to use it. Here is a way to complete the task:

> Now that you have the Weather Checker on your screen, please locate a restaurant nearby. Your favorite food is Italian, and you are on a tight budget.

The task deliberately does not indicate a price range so that the respondent will personalize the task and interpret it for himself or herself, making the scenario more realistic and therefore more like how the application will perform in real life.

Once each task is completed, ask questions about the process:

- How did you feel as you attempted to complete the task?
- How could the application be made easier?
- What was the most confusing aspect of the process?

Timing and flow of your discussion guide

It is always difficult to gauge how many tasks can be completed during a one-hour interview. For this reason, **pilot tests** are an invaluable tool. A pilot test is a single interview conducted in an informal environment with a respondent. (We do not separate out the cost of the pilot test, but include it in our basic service package; it is a required part of the research process.) The respondent need not match your recruiting criteria, since your primary goal is to test the timing and flow of the discussion guide not to test the usability of the product. Set up the pilot test as you would for your planned research. Include observers, but let them know your goals – to refine your discussion guide. Be aware that the observers will still draw usability conclusions from the pilot test. It is your responsibility to urge them not to react based on one interview and to wait until the full set is conducted, usually the following day. Developing theories on the fly is commendable, but asking developers to pull an 'all-nighter' to make changes is a very bad idea.

After the pilot test, refine your tasks and order them to get coverage of the most important areas of the user interface first, pushing the less important tasks to latter portions of the interview. Do not edit them out, since if a respondent sails through the process you will want those extra tasks to learn as much as you can.

Follow-up questions

After the tasks, you can ask each respondent questions about the overall experience. Be wary of asking yes–no questions here as well. Be especially wary of the almost obligatory 'Did you find the user interface to be easy to use?'

question that your client will expect. We ask it anyway, but warn clients that respondents almost always say 'I found it really easy to use'. They say that because they know they are about to receive their incentive, and pleasing the moderator is a natural human instinct, especially when money is involved. Modifying this question to fit a Lickert scale, such as 1 = hard to use, 5 = easy to use, is better, but respondents still want to please the moderator. Respondents who fail nearly every task still say the application is really easy to use. Instead, ask respondents what applications the tested product is like. This question could yield very interesting information.

Audiovisual Setup

Although video is not required for effective usability tests, one can otherwise rely only on notes and memory in the follow-on analysis of the usability interviews. It is also impossible to construct a 'highlights tape' from memory alone. Recording your interviews with video is an effective way to capture respondent actions and comments. Note, however, that it takes the same amount of time – or longer – to watch the video as it does to watch the live interviews. When you watch the video, you are likely to need to play back sequences multiple times, the opportunity for which does not exist when you see live action. There is much less detail available from the video recordings because of technology limitations. Video recording captures only some of the visuals, and it is impossible to create a wholly natural environment and at the same time capture what the device display shows, since the video equipment always gets in the way.

For working prototypes, which we test in focus-group facilities, a sophisticated audiovisual setup is required. You may not have access to all of the equipment mentioned here, but note that costs fall all the time, and one can put together a good video setup for less than US$5,000. Types of equipment are listed in favor of specific equipment manufacturers and models. The introduction to audiovisual setup here is meant to give you an overview of what to request of a video specialist, not to provide a shopping list and how-to guide. Video configuration is complicated and frustrating, and the equipment user interfaces are very difficult to use. Usability testing is a tiny niche market for this equipment, and, although it performs admirably, the learning curve is formidable.

However, if you managed to purchase and set up your home entertainment system with five-speaker surround sound, DVD player, stereo, CD player, television, satellite dish, cable, digital video recorder ... without calling anyone for help, go ahead and give it a try.

Illustration 6.3 shows the most basic set of elements required of an audiovisual setup. The phone is mounted to a tabletop tripod with Velcro®, and a document camera is trained on the telephone's display and keypad. Not shown are omnidirectional microphone, power cables, and signal cables. Although the phone may be wireless for typical use, it needs to be powered for a full day of testing. All of the other equipment is of the standard, wired type.

In addition to a video representation of the phone, you will want a video representation of the respondent's face. For that you can use a standard camcorder. You will combine the signals from the document camera and camcorder with a video mixer. The output from the mixer, along with that of the microphone, will go into a video recorder and television monitor.

Desktop and laptop computers can be easily rigged or wired for video recording. Handhelds and phones require setups that are more conspicuous and obtrusive. Most handheld devices and phones have no ability to echo the screen display, as is common with desktop and laptop computers. Desktop and laptop video is easily split and converted into an acceptable television signal, but smaller devices require a video camera to be trained upon the device. The device must either be mounted in some fashion where the camera can record its

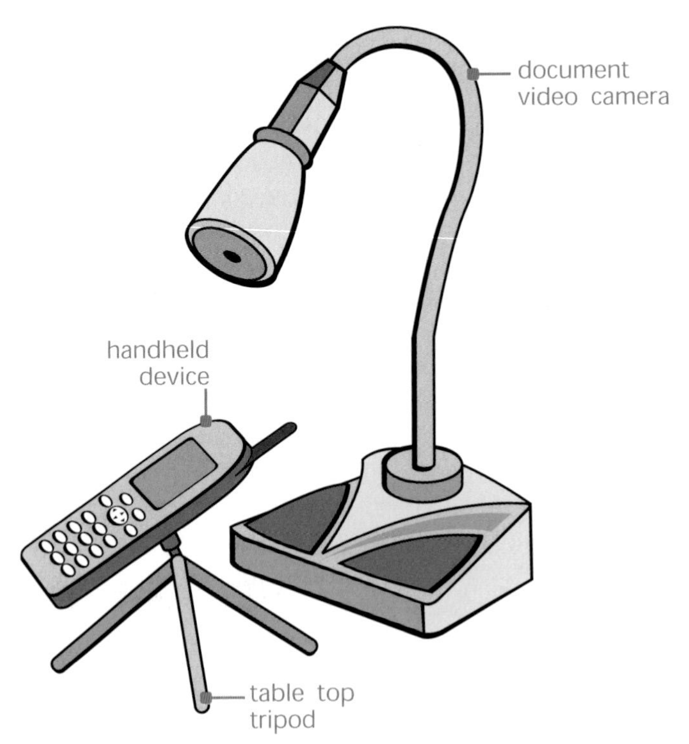

Illustration 6.3 Video setup for usability study

display, or the camera must be mounted onto the device. In the latter case, the camera must obviously be very small and lightweight. However, even small, lightweight cameras require cabling for the signal to get back to the recording device, so there are drawbacks to each approach. I feel that mounting the phone is a minor sacrifice, since the respondent is still able to grasp it and has access to the side and top buttons. The setup used for the Sprint PCS/NeoPoint 1000 Usability Study (Appendix C) is shown in Illustration 6.4.

Illustration 6.5 shows the result of the setup described above. The phone image comes from the document camera, and the respondent image comes from the camcorder. Both images are combined with the video mixer. The

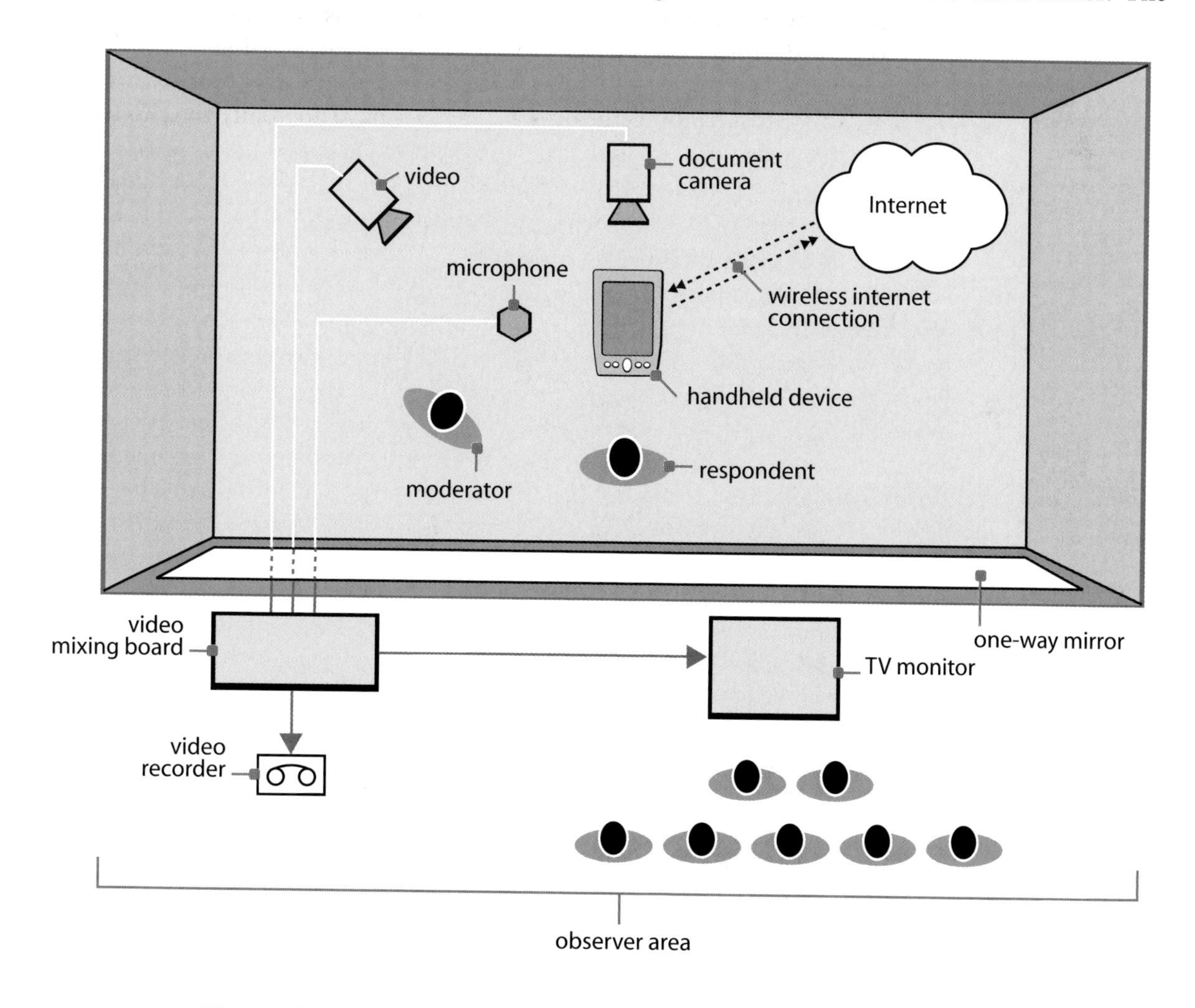

Illustration 6.4 Video setup for Sprint PCS/NeoPoint 1000 Usability Study

Illustration 6.5 Video image of phone and of respondent with phone

mixer's output, along with that of the microphone, is sent to a videorecorder and a television monitor. Note that the camcorder is not used to record video. We used a camcorder instead of a stand-alone video camera since camcorders tend to have better zoom and autofocus capabilities and they are also less expensive than stand-alone video cameras.

Digital video is superior to analog video since it can be duplicated without loss. However, since the camcorder's signal is going to the video mixer, it makes no sense to use a digital video camera. The important component is the video recorder, and there are a few good portable digital video recorders available. VHS video recorders are the worst available. DV (digital video) recorders come in several formats, all of which are adequate for this purpose. When high-definition television (HDTV) is standard, that format will be the best of all, since its resolution is better than both NTSC (National Television Standards Committee) and PAL (phase alternating line) formats, the two most common formats used in most of the world.

When selecting a facility, be sure to ask if they have a **passthrough**. A passthrough is a hole in the wall between the observer area and the respondent area where you can pass your video and audio cables. Without a passthrough, you will need to route cables through doorways and down the hall, which requires a lot of cable, gaffer tape, and time to set up.

Event logging vs. transcription

Many products are available that enable event logging during video recordings. An event logger is a combined hardware and software solution that creates a file with analyst's comments and that overlays invisible time codes on the videotape as it is being recorded. Event loggers have rich software user interfaces that allow complex tagging of events during interviews. After interviews, the analyst can replay the videotapes through the event logging system automatically to pull video clips for video highlight generation. Many researchers depend upon event loggers, but I feel that one should conduct interviews and analyze the data 'the hard way' at least a few times before moving to a more technology-intensive solution. Learn the process and understand the steps before taking shortcuts – there is plenty of technology to learn with a basic setup.

We do not use event logging software or hardware and have no intention to do so in the near future. We transcribe each interview from videotape, documenting respondent actions and commentary. The transcription process is an incredible boon to the analyst, whose knowledge of the product increases dramatically. Each one-hour videotape takes approximately four hours to transcribe, making the transcription process a three-day investment for six one-hour interviews. Three days of tortuous transcription pays a heavy dividend in insight. Our comprehensive analysis (including transcription) is far superior to our accelerated analysis (without transcription). Event logging would facilitate creating video highlights, but with accurate transcripts we can spend less than an hour identifying the best quotes and scan them into a digital video editor in less than one day.

Conduct Interviews

We typically conduct six one-hour one-on-one interviews for a project. After years of conducting six interviews, we find that each and every one of our clients has been pleased with this amount. Almost every client is at first concerned that six interviews won't be sufficient, since six data points are hardly enough for a quantitative study. However, as was said at the outset, usability testing is inherently qualitative, and after four interviews we typically observe trends. The fifth and sixth interviews confirm or deny these trends, and it is reasonable to conduct six interviews in one calendar day. Since we conduct all the interviews in one day, our clients usually attend the full suite of interviews, which we strongly encourage. Some involved tasks will require longer interviews, and when target audiences are disparate, each audience must be represented by at least four respondents. However, I feel strongly that it is far better to conduct multiple suites of usability tests with design improvements implemented in-between than doing a tour-de-force of usability tests only once in a design process.

The moderator

Interviews require a moderator who directs and supports the respondents as they attempt to accomplish tasks. The moderator's job usually does *not* include teaching the respondents how to use the product or helping them when they fail at a particular task. The most difficult thing about moderation is that the moderator must watch his or her respondents struggle without helping them. It is human instinct to help people when they have difficulty, but the moderator cannot do so and gather actionable usability data at the same time. However, in order to discover usability problems in areas that come logically after a task that the respondent is unable to perform, the moderator must then help the respondent through the trouble spot, *after documenting the respondent's failure to complete the task*. It is actually preferable *not* to help the respondents, but to keep going. Sometimes respondents will discover how to do something later in the interview; this 'a ha!' reaction is rewarding to both respondent and researcher – and client! Help respondents only when absolutely necessary to stage the user interface for the next task.

Moderators need to be nonthreatening, patient, and friendly. Business casual attire is usually preferable to suit and tie, but anything low-cut or otherwise revealing should be absolutely avoided, as should fragrance of any kind. The researcher should be approachable and not overly attractive. I wear loose-fitting, unlabeled clothing and no jewelry, although a watch and/or wedding band are certainly acceptable.

Protocol

Usability testing utilizes the **think aloud** protocol. Quite simply, respondents are encouraged to speak their thoughts. For reticent respondents, I ask them, 'What are you thinking?' Only rarely do I interview respondents who are too chatty. For these respondents, I tell them, 'In order to complete this interview on time, I will probably need to interrupt you and move you to the next task a lot. When I need to do that, I will either say "thank you" or make a "T" with my hands. Please do not be offended by this necessity'. Respondents know when they are too talkative – these people probably get less polite feedback from colleagues and friends all the time.

Avoid making jokes of any kind, since, as respondents are already nervous, they can easily be offended. Offending a respondent will reduce the quality of the interview and could cause an early end to it. If a respondent wants to leave, they will.

If respondents become invested in a task, they perform better. They will still fail if the usability is poor, but they will try alternative methods rather than quickly 'give up'. Let respondents know their goal, and begin at a logical starting point in the application. The ambiguity created by the gap between starting point and goal is your opportunity to observe the respondent as he or she solves the problem. The problem solving is the most important part of the usability interview and is the time for the moderator to be silent, simply observing and recording what the respondent does. The respondent should not be silent. Remind him or her to 'please think out loud as you complete this task'. Most people are unaccustomed to observation, and might be shy. They will need to be reminded to share their thoughts.

Success and failure

Let respondents solve problems their way, so that your application can be a useful tool, not a puzzle for them to solve.

When respondents succeed, what do you learn? Usability testing is not about validating good user interfaces; it is about identifying problems. If a respondent succeeds, give him or her a more complicated or difficult task. Do your job: break the interface by finding where and how respondents fail. If it is really hard to break, it must be good, right?

If a respondent goes down the wrong path, let him or her keep going. Failure is interesting, and so is discovery. Much of the time, a respondent will initially do something that will not contribute to success, but then will stumble upon a successful tack. If you stop the respondent as soon as he or she does something wrong, you will prevent him or her from correcting that later on. One could argue that 'stumbling' is unlike what happens in the 'real world', but if several of your respondents stumble in the same way, that approach would occur in the shipping product too. You will have found a set of behaviors that will occur in your final product.

Support alternate methods of solving problems – do *not* punish with error messages redirecting users to 'preferred' methods. When respondents get frustrated, encourage them, and be careful not to help them until you have exhausted their patience (or that of your observers).

If respondents fail in a catastrophic way, meaning that they cause data loss or shut down the product, etc., you need to step in and provide assistance. Before you provide assistance, note the task failure, then reset the stage and let the respondent try again. If it appears that the respondent is going down the same wrong path, step in and provide a hint. If the respondent then completes the task, he or she still failed, but you can test later areas of the user interface by moving the respondent through the trouble spot. We call testing areas of the interface 'coverage'. If respondents all fail in the same place, in order to cover the rest of the user interface you need to provide instructions. Sometimes I skip tasks entirely, but provide instruction to get to the later stage in the process. Observers must know what you are doing before or after you provide this level of assistance since uninformed observers may feel that you are 'leading' the respondent. You are, but only in the interest of testing the entire interface. You already know that the first part is broken; you still need to test remaining stages.

When respondents get stuck, probe them to determine what could prevent the same thing happening in the finished application:

- What are you thinking?
- What on this page might help you accomplish your goal?
- What is missing from this page?

Only after exhausting the 'why' questions that might help you design in usability should you start to offer hints to the respondent. Your hints should initially be very basic, and eventually walk the respondent through the process. Even when respondents fail, however, it is possible to learn a great deal. Ask them questions throughout, and show patience throughout this phase. They have the hardest job in the process: using your application.

Using floaters

Floaters are sometimes brought in when respondents who do not meet the recruiting requirements have somehow slipped through the recruit. If you need to dismiss a respondent, let him or her know that you need to check with your client about the performance of the 'back-end server, since it has been faulty today'. Exit the room and inform your observers about the concerns you have. Once you return, inform the respondent that 'The server went down, and we will have to terminate the interview. You will be paid. Thank you so much for your help'. Then escort the respondent out, wait for him or her to leave, and then start over with the floater. It is not necessary to upset the respondent with your real reason, since such a situation can have extremely unpleasant results. Always compensate respondents who show up for research on time, whether or not you interview them.

Observers

The observers are team members who will watch the usability interviews. Most focus-group facilities can seat up to ten observers. The more, the merrier. As long as you provide food, people will be content. Most important to include among this group are the developers, who will build the product. The developers are the final decisionmakers when it comes to usability, since, despite what anyone thinks, they control what goes in and stays out of a product. Include as many team members from each functional area as you can, since every contributor will learn something different from the research.

Encourage observers to take notes during the research, identifying usability issues, bugs, design recommendations, and feature ideas. Watching respondents while they struggle with a product is an incredible idea generator. Many of the ideas generated during usability testing get lost, but the ones captured by the observers during the interviews and shared during the debrief stand a chance of being implemented.

Debrief

The debrief session is an excellent opportunity for team members to share their thoughts about the product and how it performed during testing. It is also your best opportunity to gather input for your top-line notes and findings. Define issues first, then return to solving problems. Try to focus on issues only, since identifying issues is far quicker than solving problems. Observers will be exhausted, and your time with them is limited. Capturing all of the issues they have identified is critical during the debrief, and, once identified, solutions can be discussed later. However, if you miss issues during the debrief, they may be lost entirely.

We conduct a one-hour debrief session immediately following six consecutive one-hour interviews. The moderator facilitates the debrief session, which is held in the respondent area. The moderator uses a flip chart to list issues that the team raises, and identifies them as issues, bugs, or recommendations. The flip chart pages are later transcribed and the content is edited into the top-line notes, which are sent out the next business day. We often enhance the top-line notes with illustrations, but the key priority is getting the document out the next day, since the observers are probably already working on resolving the problems they discovered during the interviews.

Findings Documentation

The findings should be delivered in two passes, first a set of top-line notes, and second, the full findings document. The top-line notes contain all of the discoveries documented during the Debrief session, broken into product functional areas and separating out the issues, bugs, and recommendations. The cover of the document should include the build number, or release information for the product, since it is likely that this findings document will not be the last for your product. The unabridged Sprint PCS/NeoPoint 1000 Usability Study findings is included as in Appendix C.

You may not know the audience for this document ahead of time, so write it for someone who is unfamiliar with the product, as you may have been before the project started. Define project-related and company-related jargon. Your findings documentation may survive your project by months or years, so in order for it to be helpful it must clarify the product, the research intentions, and what was learned. Sections to include are:

- executive summary
- background
- goals
- methodology
- platform and test bed
- facility
- respondents
- analysis
- recommendations
- respondent data
- product images
- respondent screener
- discussion guide
- transcripts

Executive summary

The executive summary is a summation of the background, goals, platform and test bed, facility, respondents, and analysis sections. As such, it cannot be written until the very end of preparing the findings document. Optimally, it is only three or fewer pages. Be careful not to strip out any information critical to a reader's quick understanding of the tests and findings. Edit it to make sure the section holds together, and call it a day.

Background

Tell the reader what was tested, the number of respondents, where the research was conducted, and describe the document. Explain how the document is organized and discuss how the document may be used by the reader. Assume the reader has only a basic understanding of the product that was tested and little or no knowledge of the process of usability testing.

Goals

Simply put, this section explains why you did this project. The goal was not simply to test the product's usability, although your instinct may tell you just that. You tested the usability of specific areas or features of the product, so articulate those areas and features as goals. The research may also have tested the target audience definition, since it may be the first time that the intended audience was presented with the product. A project goal may even have been to identify audience response to the product.

Methodology

Describe the approach that you took for the research. Detail the structure of the interviews, how long each of them was, and the protocols used, such as 'think aloud'. Describe at a high level how the audiovisual setup was organized. Do not justify your efforts; merely describing what you did is sufficient.

Platform and test bed

Describe the equipment or prototype used for testing. Catalog the relevant issues, such as bandwidth, display type and size, input mechanism, any mounting necessary to secure the product, etc. Let the reader know what you did, and a bit of why you did it that way. Describe problems encountered during the research in your analysis, not here. This section is intended to tell your reader what you started with, not the challenges you faced along the way.

Facility

Describe where the tests were conducted – not just the location, but what type of location. Was it a focus-group facility, a dedicated usability lab, a conference room? If the room had special features such as a one-way mirror or sophisticated networking capabilities, describe the features you used here.

Respondents

Write a qualitative description of the participants in the study. Refer to the respondent screener, which is included later in the findings document. This section is a good place to discuss any insights about the respondent pool you or the team had, or to document any problems that are necessary to share with readers.

Analysis

Analysis of usability data results from the qualitative measures of respondents' performance. In our analysis we endeavor to prove each point with quotes from the respondents. The analysis is *not* our evaluation of the product, since usability testing is *not* heuristic analysis. Heuristic analysis is an expert review of a product, while usability testing, as you will recall, is the objective study of respondents while they attempt to perform tasks that involve finding information or performing transactions.

One of the two most important sections of the document, the analysis is broken down by functional area of the product. The discussion guide will often guide you in how to write this section, since most tasks focus on a single product function. As a result of years of client input we write our analyses with summary paragraphs and illustrating bullet points. We illustrate findings with diagrams, product images, and respondent quotes. The quotes prove our points.

The first paragraphs of the analysis describe the overall impressions you gained during the interviews and also describe the material covered. As the discussion guide is a *guide*, and not a script, you likely rearranged the tasks and covered additional ground. Some tasks may have been skipped. You may feel that it is necessary to share this information, but leave it out if you do not feel that it will benefit the reader.

For each functional area of the product that you tested, write what you learned, both good and bad. Describe usability problems, and support the problem descriptions with evidence from your notes. Leave out judgments and suggestions – keep the analysis as objective as you can. Do not detract from the credibility of your research by muddying the analysis with recommendations.

Recommendations

We tend to write a recommendation for each point made in the analysis. Recommendations are direct, brief, and instructive. Recommendations are *not* specifications for product features. If you are producing product specifications, your findings document will take too long to produce. Get the Findings out, and then move on to redesign. There are also many instances in which you will not get the opportunity to do the redesign yourself; the recommendations are a to-do list of what designs need to be revisited, with your viewpoint of how they should be addressed. Your recommendations are more likely to be followed if they suggest rather than demand. Words such as 'must' are best replaced with 'should', or, better yet, 'should consider'. The forcefulness of your recommendations should be less obvious than their reasonableness and quality.

Respondent data

This is a summary of whom you interviewed, typically in spreadsheet form. Include the dates and times that interviews were conducted and how respondents were compensated. The discussion of the respondents occurs earlier in the document, so keep the information here quantitative.

Product images

Since the product will change soon after your usability tests, make sure to grab images quickly and store them for later comparison. Capture every screen tested and images of the hardware, if relevant. The individual snapshots in the printed version of the document might be quite small, so your digital archive may be more valuable. Press a CD if necessary.

Respondent screener

Include the exact questionnaire used to recruit participants, formatted to fit within your findings document.

Discussion guide

Include the exact discussion guide used by the moderator during interviews, formatted to fit within your findings document. As all usability test discussion guide documents include more material than can be covered during the interviews, it is not necessary for you to comment within this section about the material. Use the analysis section to provide that type of insight.

Transcripts

Transcripts are extremely helpful for producing the analysis and the video highlights. Most researchers do not include transcripts of sessions. Our analysts have repeatedly stated that the process of transcription is extremely helpful in teasing out the most relevant analysis points. Quotes from the transcripts provide obvious proof of these points, from the respondents' perspective.

Organize transcripts by task rather than by respondent. Identify task times from your video, so that you can later produce video highlights easily and so that interested readers can cue unedited video and watch for themselves.

Video Highlights

As we discussed in the audiovisual setup section, the production of video highlights is an excellent way to encapsulate your findings (Illustration 6.6). Seeing respondents fail at tasks – and hearing what they have to say about the product – can have an incredible impact on your client. The process of producing video highlights is time-consuming but can be an eye-opener, particularly for managers. These executives might attend testing, but are generally unable to observe an entire day of interviews. Video highlights also serve as an

Nomenclature

Confusing Soft Key labels

Unfamiliar menu items

Initial Caps vs. ALL CAPS

Illustration 6.6 Video title page (left) and video clip of respondent, including respondent identifier (right)

archive of the user experience, which can help drive feature design for several months following the customer release of your product.

We start with completed transcripts, which include video times. We identify the best quotes and then scan them into a digital video editor. Within the editor, we organize the clips into content areas that match the functional areas of the product, and create title pages with bullet points describing the clips contained within each 'section'. We typically order the clips within each section chronologically, and title each clip with section information and the respondent number (1–6 in our case).

The video highlights end up being a 15–30 minute documentary about the usability issues encountered, with clips proving the points made in the findings document. Be aware that many viewers of the video will not read the findings, so provide enough information to enable them to make design decisions.

Prioritize Issues

The manager responsible for the release next needs to decide – with the team's help – what gets fixed and when. The findings section provides a list of action items to be addressed by the team members responsible for making decisions about scheduling and the marketability of the product. This team should meet to go over the issues raised and discuss which ones will get addressed, according to what schedule. That meeting's result will be an action plan, which will drive the next round of design and development.

Solving the design problems identified during usability testing is not a quick task. The participatory design process, prototyping, and further user testing are the official way to solve these problems. However, many usability issues can be remedied with terminology changes and simple reordering of interaction elements.

Conclusions

Usability testing is not a check-off item; it should be part of the iterative design–development cycle. Test usability during the design phase and test the product when it is nearly complete as well. After launch, there are more options, which are most effectively deployed in combination. The customer support team will have strong opinions based on the calls that they receive. Log files of user activity will indicate where visitors spend their time and how frequently they return. Marketing will conduct site visits and focus groups to gauge customer opinions. In concert, all of these research methods will enable your marketing team to target specific areas for usability enhancement in subsequent versions of the product. After your product has matured, perhaps for six months or a year, usability testing can again be of tremendous value. The product is fundamentally usable, since you followed good iterative design practices, but you no longer have a keen idea of what the user experience is like for a newcomer. Usability testing of a released product provides a richness of detail that prototype testing cannot.

Usability testing tells you not only about the problems; but also can spark ideas for new features and product streamlining. The results of a comprehensive usability study, with transcripts and video highlights, can effectively communicate the product's user experience to executive management, who may otherwise be unable to witness first hand the product in use by its audience. Accelerated usability research conducted during the design phase does not offer this type of rich documentation, since its goals were to help refine the design rather than to persuade a company's decisionmakers that change was necessary.

Summary

Usability testing is the objective study of members of your target audience as they attempt to use your product. Usability testing provides a qualitative measure of your product's ease of use. Test early in a project cycle, and test again prior to release. Six interviews is a reasonable number, since trends begin to emerge after the fourth interview, and six interviews can be conducted in one day. Most usability studies are conducted with one interview participant (called a respondent) at a time. Formal usability tests are conducted in dedicated labs that are outfitted with one-way mirrors, but many usability tests can be conducted in an ordinary conference room. Labs allow observers to watch without interfering with the interview itself.

Paper prototype usability tests are conducted differently from those of live prototypes. Tests using paper mockups are often conducted with two respondents, called 'co-discovery'. Paper prototype pages are dynamically created by team members based on where respondents 'click' or 'type'. Functional specifications are an effective way to deliver the results of paper prototype usability tests, while findings documents are the best deliverable from traditional usability research.

The usability testing process starts with writing a respondent screener (participant questionnaire). While participants are being recruited, the discussion guide is written. It is the list of questions read by the moderator during the interviews. The interviews are conducted next. A debrief session is conducted immediately following the interviews. Top-line notes get sent out the next day, followed by the full findings. Video highlights are an informative companion to the findings documentation.

Appendix A
Handheld History

A Brief History of Desktop and Handheld Devices

Following is a timeline of technical and political progress that led to the development and adoption of handheld devices. I have set in bold 14 key points in the timeline, which mark significant leaps in this history. Sources are tagged in square brackets at the end of entries. A source listing is provided at the end of this appendix.

1842 — The first fax machine patent was issued to Scottish inventor Alexander Bain. It read text written in raised metal letters and transmitted it through telegraph lines [1].

1865 — A commercial fax system was established in 1865 by Italian Giovanni Caselli between Lyon and Paris in France.

1876 — **Wired (telephone) communication was invented.** Alexander Graham Bell invented the first telephone. Thomas Watson built it [2].

1902 — An optical scanner was added to the fax machine in order to establish fax service for newspaper photos.

1910 — Lars Mangus Ericsson created and carted around the first car telephone. It was not wireless – Ericsson's wife Hilda hooked a contraption made of two sticks with a wire between over the nearest telephone wires, patching in to an operator.

1921 **Radio communication was invented.** The Detroit (MI, USA) Police Department used the first pager-like system [3].

1927 **Government involvement in radio communication began.** The US FRC (Federal Radio Commission) was formed. It was the precursor to the FCC (Federal Communications Commission) [2].

1934 Congress created the FCC. The FCC regulated landline phones and the radio spectrum. 'The federal government gave the FCC a broad public interest mandate, telling it to grant licenses if it was in the "public interest, convenience, and necessity" to do so. The FCC would now decide who would get what frequencies' [2].

1937–39 **The first two-way radio phones were in use.** Large 'radio phones' were already in use in Holland, although they may only have been two-way radios employed by mariners [2].

1938 **Handwriting recognition was invented.** George Hansel was issued US Patent 2,143,875 for machine recognition of handwriting [4].

1946 **The first mobile telephone service appeared.** The first commercial Telephone Radio Service appeared in 1946, for vehicle radios [2].

The first circuit boards became commercially available [2].

AT&T and Southwestern Bell introduced the commercial mobile radio-telephone to private customers. Vehicle radio-telephone licenses were granted by the FCC [2].

1947 D. H. Ring of Bell Labs wrote a technical memorandum detailing the first known account of what cellular telecommunications would be like [2].

Bell Systems asks the FCC to free up more mobile radio channels.

1948 **The transistor was invented, enabling electronics to miniaturize.**

Automatic radiotelephone service began in Richmond, IN. No operator was needed for most calls. It was developed by the Richmond Radiotelephone Company. Most of the manual systems remained in service until 1960 [2].

Bell Systems introduced the transistor, having wide implications for miniaturization of radio equipment. Most radios would still rely on tubes until the late 1950s.

Western Union started its desk fax service [1].

1940s–1950s The FCC became a conservative agent backed by and for 'Big Industry'. It developed an especially close relationship to the broadcasting industry, especially RCA. The FCC suppressed FM development 'for decades'. 'The FCC designated no private or individual radio-telephone channels until after World War II' [2]. It devoted the most space to the agencies it thought would aid the most people, including emergency services, utility companies, government, etc.

The FCC, Federal Trade Commission, and Justice department were all involved in allowing and curtailing the reach and influence of AT&T. The FCC allocated a few more channels in 1949, but gave half to other companies wanting to sell mobile telephone services. These Radio Common Carriers, or RCCs, were the first FCC-created competition for the Bell system. Their activities may have halted development of cellular for 10 or 20 years [2].

1954 **The first commercial production of transistors occurred.** Texas Instruments started commercial production of silicon-based transistors, allowing for greater output with less power. This led to further miniaturization of radio components. The first commercial transistor radio was powered by TI silicon transistors.

1955 Motorola invented the 'Handie-Talkie' tone-only pager (Illustration A.1). Users would hear a tone and had to call an operator to retrieve their message [5].

1956 AT&T and the Justice department settled an anti-monopoly lawsuit. Bell Labs stopped supplying radio equipment to private and public concerns [2].

Illustration A.1 Motorola Handie Talkie. Reproduced with permission from Motorola Inc.

	Motorola produced the first commercial transistorized automobile radio [2].
1958	The RCCs (Radio Common Carriers) continued to make improvements to their mobile systems, including automatic direct mobile-to-mobile calls [2].
	TI's Jack Kilby invented the integrated circuit [2].
1962–1969	The Internet was conceived, while the Arpanet grew [6].
1966	Xerox introduced the first general purpose fax machine that used standard phone lines [1].
1972	Hewlett-Packard introduced the HP-35, the first handheld scientific calculator [7].
1973	Dr Martin Cooper filed a patent for Motorola entitled 'Radio telephone system' [2].
	Email quickly became the most popular application for the Arpanet [6].
1976	Only 545 customers in New York City had Bell Systems Mobile Phones. Another 3700 customers were on the waiting

Illustration A.2 The Apple II. Reproduced with permission from Apple Computer Inc.

list; 44 000 people in the USA had AT&T mobiles, with 20 000 on a waiting list of 5 to 10 years [2].

1977 **Apple Computer released the Apple II, the world's first personal computer** (Illustration A.2).

1980 **Radio Shack introduced the TRS-80 Pocket Computer, the first handheld computer.**

1981 Xerox introduced the Star, the world's first computer with a bitmapped display, windows, icons, and a mouse.

The first North American Cellular System in Mexico City, just one 'cell', was started [2].

The Nordic Mobile Telephone System introduced Europe's first cellular service (NMT450) in Denmark, Sweden, Finland,

Illustration A.3 The IBM personal computer. Reproduced with permission from IBM

and Norway. Operating in the 450 MHz range, it was the first multinational cellular system [2].

IBM released the first IBM PC (Illustration A.3).

1982 Development of GSM, the Global System for Mobile communications, began with 26 national European telephone companies [2].

The term 'Internet' was used for the first time [6].

1983 Apple released the Lisa 2/10, Apple's first computer with a graphical user interface. The Lisa was a commercial failure. The Macintosh was based on the Lisa [8].

The regional Bell operating company Ameritech began the first US commercial cellular service in Chicago, Illinois [2, 9].

1984 **Apple released the first Macintosh, the first commercially successful computer with a graphical user interface** [10].

Psion launched the world's first volume-produced handheld computer, the Psion Organizer [11, 12].

1987 John Sculley conceived the personal digital assistant. The Knowledge Navigator was a concept that Apple eventually developed into the Newton.

Canon introduced the first plain-paper fax machine [13].

1988 Japan's Ministry of Posts and Communications broke up NTT's (Nippon Telephone and Telegraph) monopoly of mobile phone systems [2].

1989 The GRiDPAD debuted, the first 'handheld', although the description is a bit of a stretch: the device was 9" × 12" × 1.4" and ran DOS on an 8088 processor. The device was too large to be accurately called a handheld. It sported a 640 × 400 display and had either 256 or 512 kb battery-backed RAM cards. Text entry was through a stylus using a character recognition engine, or by cable-connected detachable keyboard [14, 15].

1989 The European Telecommunications Standardization Institute (ETSI) became responsible for the further development of GSM. ETSI is a not-for-profit organization whose mission is to produce telecommunications standards in Europe [16].

1990 Wide-area paging became popular; 22 million pagers were in use [3].

North American Cellular Network went digital. It was backward compatible with analog phones (hence dual-mode phones) and this move, along with other small 'tweaks' such as TDMA (time division multiple access) allowed for much greater call capacity.

1991 Apple Computer petitioned the FCC to allocate a 40 MHz wide band of frequencies to use with its personal digital assistants [7].

GSM networks started operating in Europe [2].

Japan implemented the Personal Digital Cellular (PDC) standard, using TDMA [2].

The US NSF (National Science Foundation) lifted restrictions

Illustration A.4 Apple Newton MessagePad 2100. Reproduced with permission from Apple Computer Inc.

on the commercial use of NSFNET, the backbone of the Internet, opening the way for e-commerce on the Internet [6].

1992 Apple Computer Chairman John Sculley coined the acronym 'PDA' [7].

The first version of Mosaic – the first Web browser – was written at the US National Center for Supercomputing Applications (NCSA) [17].

1993 The Apple Newton (Illustration A.4) was released.

Newton was introduced at the Consumer Electronics Show [7].

Casio and Tandy introduced the Zoomer (Illustration A.5) [7].

By 1993 wireless connectivity was available for the Newton through NewtonMail Software, connected by a Motorola Digital Personal Communicator flip-phone and Motorola's Portable Cellular Connection (PCC) interface. The PCC was a small, inexpensive battery-powered box about the size of a flip-phone that connected to standard RJ11 devices, such as a Newton or PowerBook fax modem [18].

The Mosaic Web browser became available [6].

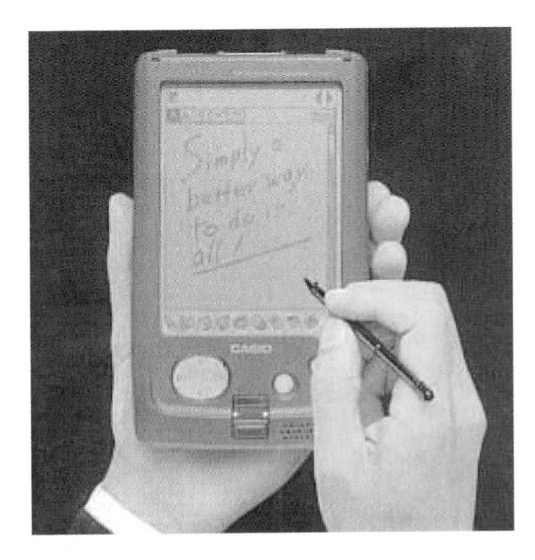

Illustration A.5 Casio Zoomer. Reproduced with permission from Casio Computer Ltd.

Illustration A.6 Sony Magic Link: the machine that used the MagicCap operating system. Used by permission of Sony Electronics Inc.

Use of the Internet grew at a rate of 341 634% in 1993 [6].

1994

General Magic released Magic Cap (see Illustration A.6). Magic Cap was a feature-rich graphical operating system for handheld devices. It relied on the ‘office’ as a metaphor, although it used many aspects of the desktop metaphor in its design. Motorola released a product called the Envoy that had wireless data connectivity [19].

Libris was formed by Alain Rossman. Libris was later renamed Unwired Planet, then Phone.com.

The Japanese mobile market was deregulated, creating the mobile ‘craze’ [20].

Ericsson started work on Bluetooth™ [20].

Netscape was formed by the original developers of the Mosaic browser [17].

1994–1997

The FCC started auctioning additional bandwidth in higher frequencies using IS-95 (CDMA, code division multiple access) and IS-136 (TDMA, in a new frequency). GSM (Global System for Mobile communication) was also added although it was much slower than in Europe, because it was placed at a much higher frequency. Many carriers introduced multimode phones that used whichever band was best at the time [2].

1995 Alain Rossman created a computer-generated prototype to demonstrate transmitting Internet data over a wireless phone.

1996 **3Com debuted the Palm™ Pilot handheld** [7].

Microsoft announced Windows CE 'Consumer Electronics' [7].

Compaq, NEC, Hitachi and Casio shipped handheld computers using Windows CE.

Motorola introduced the 3.1-ounce StarTAC mobile telephone.

Motorola introduced the first two-way pager, the PageWriter 2000. It was able to send pages to other two-way pagers and had basic PIM (personal information manager) functionality, but did not 'synch' with desktop computers [5].

1997 WAP Forum was founded by Phone.com, Ericsson, Motorola, and Nokia.

1998 Apple Computer Inc. announced that it would discontinue further development of the Newton® operating system and Newton OS-based products, including the MessagePad® 2100 and eMate® 300 [21].

Symbian was formed as an independent company by Ericsson, Motorola, Nokia, and Psion [11].

The Palm™ III handheld was introduced [7].

PDA market share: 3Com, 59%; Sharp, 17%; Hewlett-Packard, 13% [7].

The Bluetooth Special Interest Group (SIG), a primarily volunteer forum for companies to work together using short-range wireless technologies to solve customer problems, was founded. The SIG promoters include 3Com, Agere, Ericsson, IBM, Intel, Microsoft, Motorola, Nokia, and Toshiba [22].

2000 First Bluetooth products hit the market [22].

Phone.com and Software.com merged to form Openwave, maker of the most popular WAP browser.

2001 Psion abandoned its Bluetooth division and stopped producing handheld computers [11].

History References

1. Saunders, Fenella, 2001, *Discover* **22**(5), www.discover.com/may_01/breakinvented.html.
2. Farley, Tom, undated. 'Mobile Telephone History', www.TelecomeWriting.com, retrieved 9 November 2001, from www.privateline.com/PCS/history.htm.
3. www.phonewarehouse.com/pager.htm
4. EROLS (1992) users.erols.com/rwservices/pens/penhist.html.
5. Hardin, Christi, representative of the Motorola Media Intelligence Center, email communication.
6. www.pbs.org/internet/timeline/.
7. www.islandnet.com/~kpolsson/handheld/.
8. Tom, Carlson, 2001, 'Obsolete Computer Museum', Williamsburg, VA, www.obsoletecomputermuseum.org.
9. wearables.www.media.mit.edu/projects/wearables/timeline.html#1966a.
10. Sanford, Glen, www.apple-history.com, copyright 1996–2000.
11. news.bbc.co.uk/hi/english/business/newsid_1434000/1434369.stm.
12. ww6.investorrelations.co.uk/psion/custom/Corporatebackgroundpost11July2001.pdf.
13. www.hffax.de/History/faxhistory.html.
14. www.haggle.com/pen.html.
15. www.pencomputing.com/TabletPC/pen_history_dom.html.
16. ETSI (European Telecommunications and Standardization Institute) www.etsi.org.
17. www.wunderland.com/WTS/Jake/CubeArt/Descriptions/MosaicHistory.html.

18. MacNeill, David, www.pencomputing.com/Newton/NewtonNotes1.html.
19. Pen Computing, 2001, 'Magic Cap/DataRover Online', www.pencomputing.com/magic_cap/.
20. www.cnn.com/2000/TECH/computing/09/01/bluetooth/.
21. www.apple.com/ca/press/9802/NewtonDisco.html.
22. www.ericsson.com/bluetooth/companyove/history-bl/.

Further Reading

Johnson, Jeff *et al.*, 1989, 'The Xerox "Star"': a retrospective, www.geocities.com/SiliconValley/Office/7101/retrospect/.

Linden, Patrik. 1997, Contact, 'The Ugly Duckling that Became a Swan', retrieved 12 November 2001 from, www.ericsson.com/about/publications/kon_con/contact/cont15_97/c15_12.html.

PenComputing, www.pencomputing.com/.

www.ericsson.com/bluetooth/bluetoothf/.

www.motorola.com/MIMS/MSPG/Special/explain_paging/pch2.html.

Appendix B

Paper Prototyping Applications for the Palm OS®

Steps to Create a Paper Prototype

The steps to create a paper prototype of a handheld device are universal. You may not need to prototype a Palm OS® user interface, but the steps below can be applied to Windows CE, EPOC, WAP, or any other device.

1. Identify a scenario, such as:

   Joe creates an entry in his Palm OS Address Book for Scott Weiss, a business associate.

2. Convert it into a task. Here is what I wrote:

   What you see in front of you is the Palm OS Address Book.
   Please enter a colleague of yours, A. Person, into the Address Book.
   His information is as follows:
      A. Person
      The Company
      Office Phone: 12345
      Email: aperson@thecompany.com
   If respondent does not do so on his or her own:
   Please place Scott Weiss into the 'Business' category.

3. Map the pages for this task. In this case, I captured pages and screen elements directly from the Palm (see Illustration B.1).

4. Create your blinder (Illustration B.2). I scanned in the Palm™ III handheld

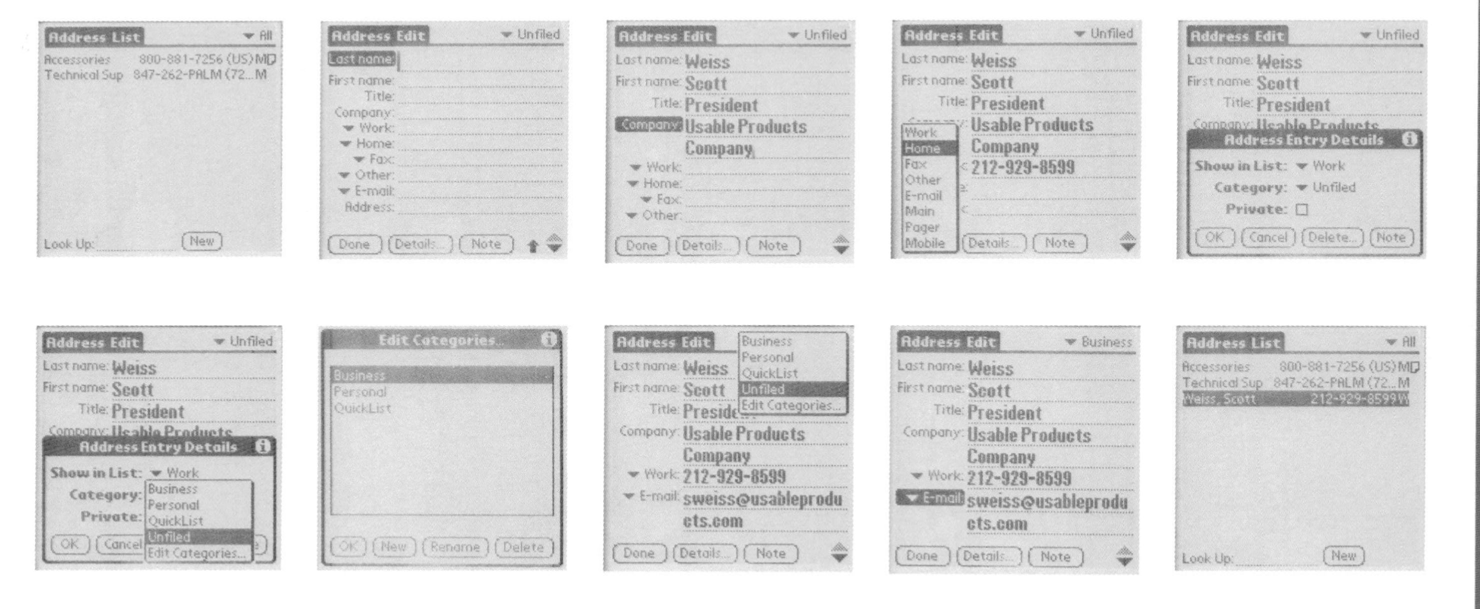

Illustration B.1 Screen shots from Palm OS® Address Book. Top row, from left to right: start state, new, 'New' clicked, partial entry, 'Home' clicked, 'Details' clicked. Bottom row, from left to right: 'Unfiled' clicked, 'Edit Categories', 'Unfiled' clicked, completed entry, 'Done' clicked. Reproduced with permission from Palm Inc.

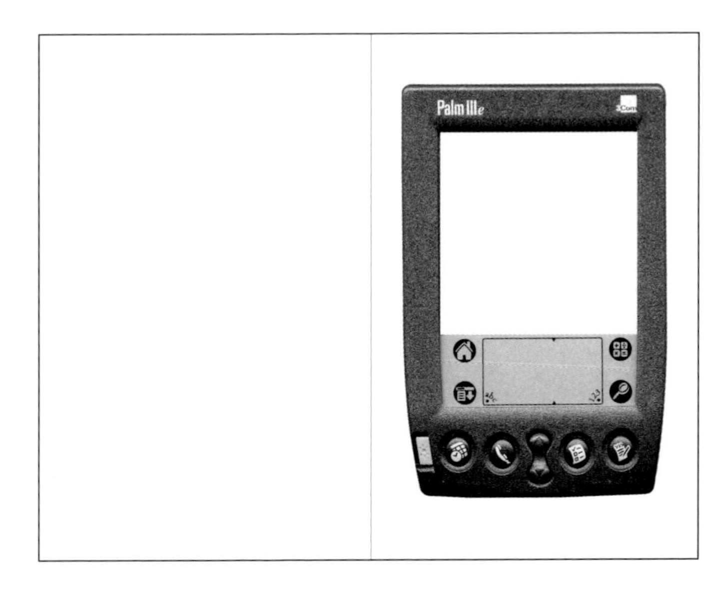

Illustration B.2 Blinder for the Palm OS® platform. Reproduced with permission from Palm Inc.

and scaled it up by 150%. I positioned it on an 8.5" × 11" sheet of card stock on the right-hand side so that the page could be folded. I then cut out the display area with a penknife. The vertical line in the center of the page is a guide for folding the blinder in half. I could easily have drawn the hardware onto the card, but in this case it was easy to scan it in.

5. Create page grids. In preparing the prototype grids for this section, I started with a drawing program, creating the perfect Palm OS page grids. Then I had an inspiration, and placed a sheet of card stock in the Palm blinder and traced inside the display cutout.

6. Next, create your pages and page elements. It is not necessary to include every detail and data item in a prototype. Note the differences between the two versions of the Address List page (Illustration B.3). I did not attempt to inverse-highlight the 'Address List' title, mainly because the task's focus is for a subsequent page, requiring the respondent only to click the 'New' button.

   I placed a piece of clear plastic on top of the 'Look Up:' text entry field (Illustration B.4). The text can be written with a wet-erase transparency marker during usability interviews. [Note: dry erase markers, though they would seem the logical choice for such an application, are permanent on transparencies. They are erasable only on glass and porcelain surfaces, such

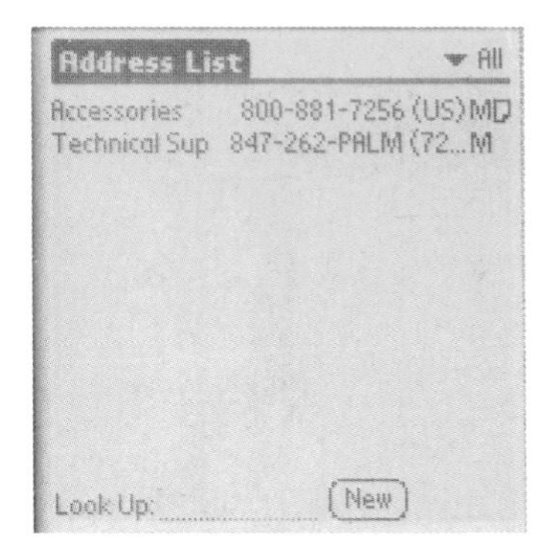

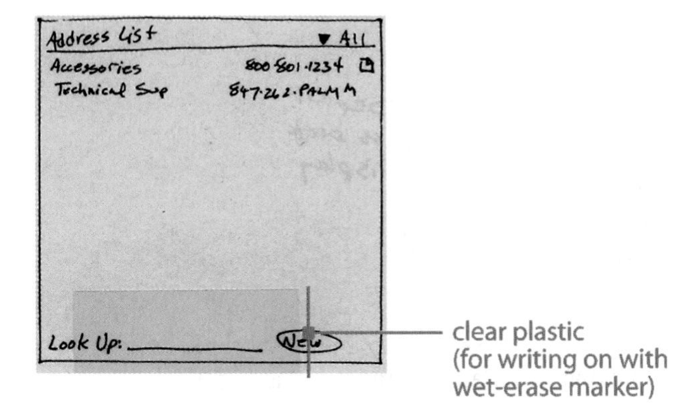

Illustration B.3 Palm OS® Address List page: actual screen (left) and drawn page for paper prototype (right). Screen shot reproduced with permission from Palm Inc.

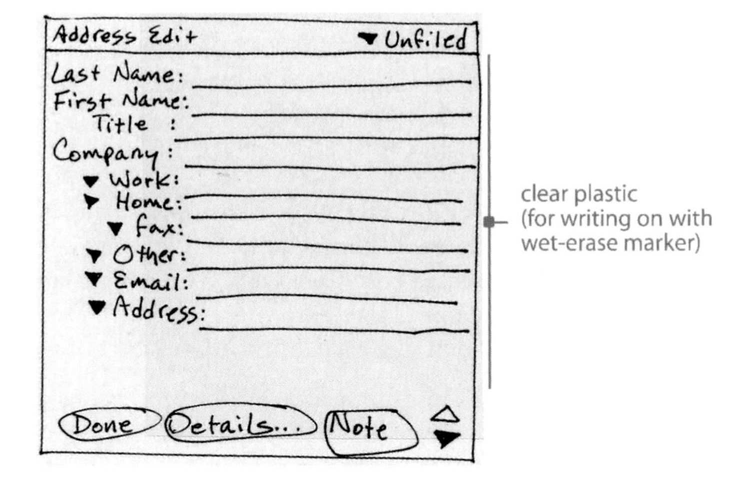

Illustration B.4 Palm OS® paper prototype Address Edit page

as whiteboards. Wet-erase markers are readily available from every office supply store.]

For each element of the prototype user interface, you will need to do the following things:

(a) Label the back with date and description (Illustration B.5).

(b) Apply restickable glue. Be sparing, as this adhesive can become messy.

Illustration B.5 Palm OS® paper prototype: Category popup list (left), information on back of Category page (middle), and Category menu items (right)

(c) Dog-ear a corner so that you can lift and stick the element on the prototype page.

Illustration B.5 shows the Category popup menu, front and back. Above at the right of this illustration is a set of each of the menu items. I included the menu triangle graphic to make it more obvious what the element is and to make it easier to place on the prototype page.

The 'Edit Categories' dialog is unnecessary for the 'Enter a colleague' task. However, a respondent might click the 'Edit Categories...' menu item, so I created a quick prototype page of it. I saved time by 'greeking' the text on the dialog and leaving out most of the dialog's functionality (see Illustration B.6).

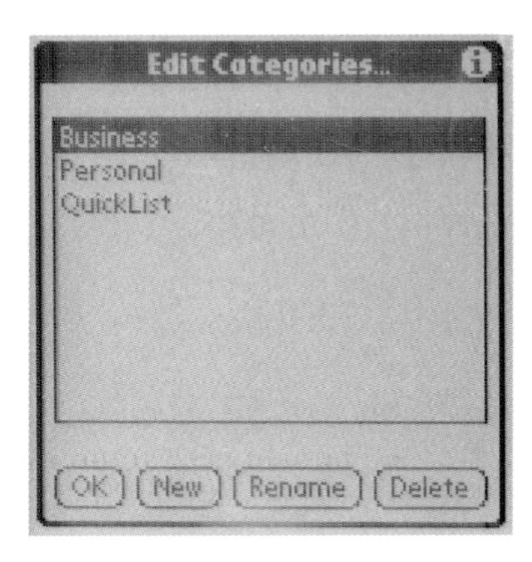

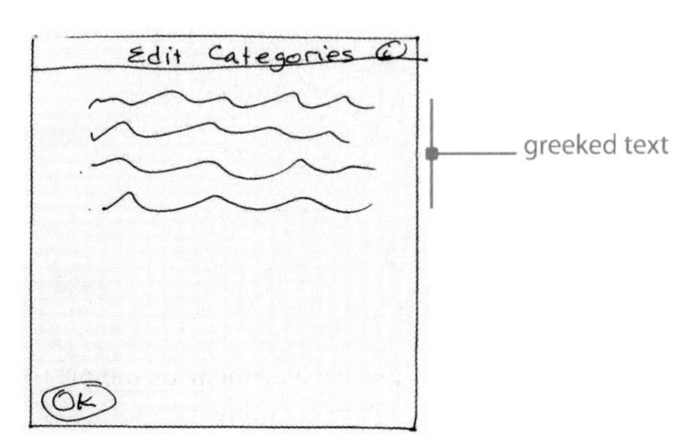

Illustration B.6 Palm OS® Edit Categories dialog: actual screen (left) and drawn page for paper prototype (right), showing 'greeked' text. Screen shot reproduced with permission from Palm Inc.

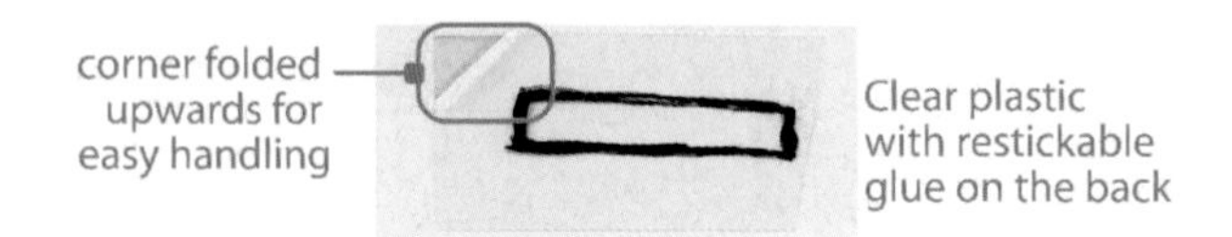

Illustration B.7 Highlight overlay

Illustration B.8 Palm OS® paper prototype Address Entry Details dialog: dialog box (left), popup menu (middle), and check box (right)

A clear plastic highlight is placed on top of a field label when its adjacent text entry field is clicked (Illustration B.7). The 'Address Entry' dialog box prototype, with the Location popup menu and check box are shown in Illustration B.8.

7. Finally, one obtains a fully-assembled page, as shown in Illustration B.9.

Illustration B.9 Fully realized paper prototype of the Palm OS® Address Book application. Image of blinder reproduced with permission of Palm Inc.

Appendix C

Sprint PCS/NeoPoint 1000 Usability Study

Respondent Screener Sample

Introduction

This section contains the actual respondent screener used for the Sprint PCS/NeoPoint 1000 Usability Study conducted in late 2000 by Usable Products Company. A respondent screener is the questionnaire used during telephone interviews to recruit usability test participants. For more information, see the section on Write a Respondent Screener in the chapter on Usability Testing (pages 161–165).

The Respondent Screener

Respondent Name: ______________________________

Telephone: ______________________________

City of Residence: ______________________________

Thank you for your time today. We would like to test functions of wireless websites and need your help. Do you have a few moments to answer some questions about your use of the web?

No	Reschedule or Terminate
Yes	☐

1a. What type of work do you do?

Web design or other Internet-related	Terminate
Cell phone industry	Terminate
Market research	Terminate
Computer programming	Terminate
Other	______________

1b. Do you own or use a mobile phone?

No	Terminate
Yes	☐

1c. Have you ever accessed the Internet through a cell phone, Palm Pilot, or similar device?

No	☐
Yes	______________

2. Have you ever participated in a one-on-one usability interview or focus group?

Yes	☐	Skip to 4
No	☐	Skip to 5

3. When did you last participate in a usability interview or focus group?

Less than 6 months	Terminate
More than 6 months ago	☐

4. Gender

Male	☐	Recruit a mix 3 Males 3 Females
Female	☐	

5. What is your age?

Younger than 21	☐	Recruit a mix
21–32	☐	
33–44	☐	
45–56	☐	
57 and older	Terminate	

6. What is the highest level of education you completed?

Did not complete high school	Terminate	
Completed high school, but no college	☐	Recruit a mix
Started but did not complete my associates degree	☐	
Have an associates degree	☐	
Started but did not complete college	☐	
Completed college	☐	
Started but did not complete graduate school	☐	
Completed graduate school	☐	

7. How many hours do you spend browsing websites and otherwise using the Internet per week *not including for email or for online coursework (school)*?

Less than 2 hours	☐	Recruit a mix
2–4 hours	☐	
5–6 hours	☐	
7–8 hours	☐	
More than 8 hours	Terminate	

8. What does the 'Back' button on a web browser do?
Note to recruiter: respondents must understand the concept of a browser's 'Back' button.

Don't Know	Terminate
Takes me to the previous web page	☐
Other	______________________

9. What is your household income?

I am a student	☐	Skip to Scheduling
Less than $25 000 per year	☐	Recruit a mix
$25 000–$40 000	☐	
$40 000–$55 000	☐	
$55 000–$70 000	☐	
$70 000–$85 000	☐	
Over a $85 000 per year	☐	

We are scheduling one-on-one interviews this month. The interview will last one hour and you will be videotaped. You will be seated in front of a mobile phone with a moderator who will ask you to find information using a wireless website.

Would you be interested in participating in the usability interview? You will be paid $50 for your time.

Please note: we require that you sign a release permitting us to record the interview on videotape.

The location is (company name, with directions).

Please bring a government issued photo ID with you. Please remember your glasses if you need them to read a mobile telephone display.

Discussion Guide Sample

Introduction

This section contains the actual discussion guide used for the Sprint PCS/NeoPoint 1000 Usability Study conducted in late 2000 by Usable Products Company. A discussion guide is the material used by the moderator during a usability interview. For more information, see the section Write a Discussion Guide in the chapter on Usability Testing (pages 168–173).

The Discussion Guide

Usability tasks for wireless web research study

Prepared by David Kevil, Richard Martin, and Scott Weiss, November 2000.

Research date/time

- November, 17–21 2000

Research location

Directions: (detailed directions follow).

Platform

- Sprint NP1000 cellular telephone.
- Dimensions: 5.5" × 2.0" × 1.0" × 6.4 oz.
- Eleven lines of text and icons with variable type sizes and proportional type.
- Tegic T9 text input, in addition to triple-tap alpha, symbol, and numeric entry.
- Sprint PCS Wireless Web browser-enabled.

Test strategy

- Each test is 60 minutes.
- Each test is with one moderator and one respondent.

Pretest statement

The following statement will be read to each respondent verbatim at the start of each interview.

We are evaluating the service and *not* you in this exercise.

You may pause or end the interview at any time if you are feeling uncomfortable.

Ask any questions that come to mind, but, due to the research nature of this interview, I may not be able to answer some of them until the end of the meeting.

Please note that this meeting is being video and audio recorded.

Please think out loud. Your thoughts are very important to us.

We may be pressed for time, and I may move you on to the next task quickly. This is not a reflection on you or your performance today.

Pretest questions

These questions will provide the team with basic information about the respondents, and will help them to understand the respondents' lifestyles and points of view.

1. How long have you been using a cellular phone?

2. How many minutes do you typically spend on your cell phone per month?

3. Do you have a Palm Pilot or other PDA?

   What kind?

   How long have you had it?

4. Have you ever used a cell phone to connect to the Web or otherwise find information without making a voice call?

   How have you used the service?

Tasks

Activating the MiniBrowser

Testing ability to activate browser.

- Without pushing any buttons, how would you connect to the Internet with this phone?

Please try it now.

- What did you think about what happened?
- How could it be easier to connect to the Internet?

Investor

Please find the current stock price for the Intel Corporation. They are listed on the NASDAQ and their symbol is 'INTC'.

- What do you think about the information you found?
- How did the information you received compare with your expectations?
- What would you change about this service?

Weather task

Please locate the weather report for today in Chicago, IL.

- What do you think about the information you found?
- What would you change about this service?

Weather task 2 using T9 data entry

If respondent used ALPHA, have respondent switch to T9.

- What does 'T9' mean?

Please locate the weather for today in Denver, CO

- What do you think about the information you found?
- How could accessing this information be made easier?
- What does 'T9' mean?
- How can T9 text input help you?

Movie times

Please locate the time for the next showing for the movie *Little Nicky* using the 10014 zip code.

- What do you think about the presentation of the information?
- How do you feel about the process of locating this information?

- How could finding a movie on this service be made better?

Checking email

Please go to MSN to check your email account. The address is 'upctest1@hotmail.com'. The password is 'abcdefgh'.

- How do you feel about the text input process?
- What do you think could be made better?

Sending email

Please send an email to upctest9@hotmail.com with the following information:

Subject: Dinner Tonight?
Body: How about dinner at Sapore at 9pm?

- How do you feel about entering email on this service?

Exiting the MiniBrowser

- Without exiting the browser please tell me how you would leave?

Please end your Internet session now.

- How does this process compare to activating the browser?
- How could this feature be made better?

Follow-up questionnaire

1. How would you rate the experience of using this phone's MiniBrowser?

1	2	3	4	5
It was really difficult	I had some difficulty	Neutral	It was enjoyable	It was very enjoyable

2. How would you rate the information you found?

1	2	3	4	5
Not useful at all	Somewhat useful	Neutral	Useful	Very useful

3. What would you change about this phone?

4. What would you change about the MiniBrowser?

Findings Documentation

Introduction

This section contains the Sprint PCS/NeoPoint 1000 Usability Study conducted in late 2000 by Usable Products Company. For more information see the section on Findings Documentation in the chapter on Usability Testing (pages 182–186). The contents list and full transcripts of interviews have been omitted. As the respondent screener and discussion guide are given above, they are not reproduced again at the end of the findings documentation. Comments in square brackets were not in the original text.

Background

Usable Products Company conducted usability research from 20 October to 12 December 2000 on the Sprint PCS Wireless Web using the NeoPoint 1000 mobile telephone. Six one-hour one-on-one usability interviews were conducted in New York City.

This document contains analysis and recommendations stemming from this research as well as documents prepared to standardize and support the research process. Usability studies involve recruiting respondents from a respondent screener, interviewing the respondents using a discussion guide, and analyzing video recordings of the interviews. The respondent screener and discussion guide are included as appendices to this document (see the first two sections of *this* appendix), as well as each respondent's answers to screener questions.

Video recordings of interviews were made, with each respondent's face and upper body shown in split-screen format to the right, and the NeoPoint 1000 telephone to the left. Transcripts of the respondents' actions and statements were made and formed the basis for the Analysis and Recommendations sections of this document.

Goals

Goals of this suite of research were the following:

- To determine the overall usability of the NeoPoint 1000 mobile phone.
- To determine the overall usability of the services that are available from the 'Sprint PCS Wireless Web'.
- To publish a study documenting the state of wireless web application usability.
- To standardize the Usable Products methods for testing the usability of hand-held devices.

Methodology

Traditional usability testing methods were used in this project, with the test moderator sitting adjacent to each respondent. For the tests, the moderator interviewed each respondent for sixty minutes. The moderator asked respondents to perform tasks while an analyst recorded their actions and whether they succeeded or failed at each task, along with task completion times and error rates. The analyst and the moderator then transcribed the video recordings in order to prepare the Findings.

- Six one-on-one interviews were conducted.
- The moderator worked with the respondents directly, not separated by a one-way mirror.
- Each interview was video recorded and audio taped.
- An ELMO document camera was used to capture the display of the mobile telephone.
- A split-screen display was used to simultaneously view the respondent's face and the telephone's display.
- The analyst was present in the room with the moderator and the respondent. No other observers were present.
- The phone was affixed to a table-top tripod with Velcro™, which allowed each respondent to handle the phone, despite its being mounted.

Platform and Test Bed

The mobile phone used during usability research was the NeoPoint 1000. This phone was selected for its large display and versatile functions. The researchers decided against using a repeater to strengthen the signal to ensure connectivity, since signal strength in the facility used was acceptable.

- Sprint NP1000 cellular telephone.
- Dimensions: 5.5" × 2.0" × 1.0" × 6.4 oz.
- Eleven lines of text and icons with variable type sizes and proportional type.
- Tegic T9 text input, in addition to triple-tap alpha, symbol, and numeric entry.
- Sprint PCS Wireless Web.

Facility

Interviews were conducted in a conference room in Usable Products's offices in New York City.

- (Detailed directions provided to the offices.)
- The conference room is not split by a one-way mirror. The moderator was seated next to the respondent at a table. The analyst was seated at adjacent table, where he controlled the video and audio equipment.

Respondents

Six respondents matching a predetermined range of traits were recruited from the New York Metropolitan area. All six of the respondents were used for interviews.

Respondents were recruited by phone interview according to the respondent screener (see the first part of this appendix).

The goal was to find web-savvy and mobile-phone-savvy individuals who would provide feedback representative of the average wireless web telephone user.

- Respondents did *not* work in the fields of market research, media, new media or other Internet-related fields, or in software.
- Education ranged from high-school graduates to college graduates.
- Respondents represented inexperienced to advanced web use levels: less than 2 hours to 7–8 hours per week, excluding email and class activities.
- Three of the respondents interviewed were female and three respondents were male.
- The respondents consisted of one Caucasian male, one Asian American

male, one African American male, two Caucasian females, and one Asian American female.

- Respondents had annual household incomes ranging from 'student' to $85 000 and over.

Analysis

The NeoPoint 1000 phone and the MiniBrowser proved to be usable for respondents, who were able to complete most of the tasks. However, every product, when tested for usability, reveals many areas where improvements can be made. The findings in this section articulate the challenges faced by respondents during one-on-one interviews.

During the course of the usability interviews, respondents were asked to complete several different tasks. Issues concerning the general usability of the NeoPoint 1000 phone and the MiniBrowser are contained in this section.

Overall Impressions

Most of the respondents had difficulty accessing the NeoPoint 1000's MiniBrowser. Of the six respondents, four respondents rated the experience of using the phone's MiniBrowser a 2 or lower on a 1 to 5 Lickert scale (5 being best).

- Most respondents reported they would have made a voice call to complete some of the tasks:
  What do you think of the information you found?
  I would rather call Moviefone (R4 54:35).
- Most respondents believed they would have been able to complete the tasks more easily and efficiently if they would have been able to personalize the telephone:
  If weather was something I was after on a day-to-day basis, then I would like a feature where I could personalize the information (R2 36:39).
- Most respondents requested more instructions on how to use the MiniBrowser:
  Finding (the MiniBrowser) wasn't bad. If I had looked through the directions, it would have been easier to find. (Without the directions) I had to fumble through it (R6 14:52).

- Most of the respondents wanted an 'Internet' button to make connecting to the web easier:
  How could connecting to the Internet be made easier?
  Have an Internet button somewhere. That would be easier and faster (R1 03:08).

Key Functions

Some respondents had difficulty finding features of the phone. This caused frustration for respondents during the tasks. Once respondents learned the various functions of each key, they did not always use the keys in the correct manner. Some keys were not obvious to respondents.

- The telephone's 'Menu' and 'Back' buttons were not obvious to most respondents initially. These buttons appeared on the phone as the letters 'M' and 'B' (see Phone Diagram, (Illustration C.5, page 234) for details).
  What do you think the 'M' and 'B' buttons stand for?
  I have no idea (R2 4:27).
- There were many variations between softkey labels across services and between cards (Illustration C.1):
  On the last screen, I saw 'Go' on the bottom. (R1 indicated the RSK label area.) *And I assumed 'Go' was correlated with the arrow that was pointing next to the number. I don't know because I pressed the number.* (R1 indicated the RSK label.) *Now it says 'Next' and now I don't know if it means next for the arrow or the next for going down* (R1 24:50).
- Most of the respondents had difficulty using the 'SHIFT' key:
  Are you done entering all of your information?
  No. Usually if I press 'SHIFT' it will change to numbers. (R5 attempted to hold the 'SHIFT' button down and press '1' at the same time.) *That's not doing anything, it's just giving me quotes* (R5 43:00).
- Some of the respondents tried to use the 'CLR' key for backward navigation:

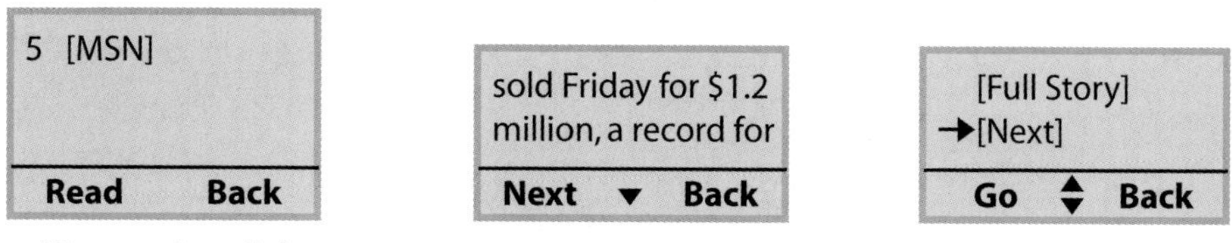

Illustration C.1 Sprint PCS/NeoPoint 1000 Usability Study: softkey variation

I pressed 'More' and now it is saying 'Select'. (R3 scrolled using the joystick.) *I'll press 'Clear' to try to get out.* (R3 pressed 'CLR'. No result.) *That's not going. I don't know what 'M' and 'B' are on the phone* (R3 3:00).

- None of the respondents used the rocker buttons on the left side of the phone. These buttons adjust volume during a voice call, and scroll the display during a data call.
- All of the respondents used the joystick with little difficulty.
- Some respondents used numbers corresponding to menu items to select. Others used the highlight/select mechanism.

Text Entry

All respondents had difficulty entering text. Text entry tasks were frustrating for respondents. Some respondents used the T9 text entry mode to complete that task. Respondents were positive about the T9 feature, but used it with varying degrees of success.

- Respondents who used the forms in Yahoo! for the Movie Task had difficulty entering the information required.
- The cursor did not move fast enough for the respondents when using triple tap:

  This is slow. When I am keying in information, the letters are not displayed fast enough so I am (tired of) typing. I have to go back and clear things, and that makes the process take a lot longer. (R4 keyed in 'upctest9' and pressed LSK.) *I got an error message that says that the email address is incomplete. Well let's try that again* (R4 15:54).
- The different text entry modes were difficult to find:

  It's not 'SHIFT', it's not 'END'. I don't want to hit 'M', because that will take me back to the Menu. ***(Moderator re-prompted.)*** *Underneath Sprint? These ambiguous keys? I don't want to hit 'ALPHA' because that will just give me question marks. I'll try it anyway just to see.* (R3 pressed the RSK.) *I guess 'SYM' means symbol. I'll hit it again.* (R3 pressed RSK.) *There is still 'OK', and now there is 'NUM', which I will assume is number. I'll select that just for the sake of it. There is the T9* (R3 25:49).
- Using T9 was not intuitive:

  What happened?

I started to spell 'dinner' and the letters kept changing. I pressed the button and the 'd' and it changed to an 'e'. This is not working. It spelled 'egg' (R2 22:55).

- Most respondents had difficulty locating the 'CLR' key located on the phone's keypad (Illustration C.2). This made it difficult for some respondents to correct typos:

  (R3 entered the email body with 'ALPHA' mode.) *Oh, there's 'Clear'!*
  What did you discover?
  'CLR' blended in with all the other numbers because it is white. I feel like such an idiot (R3 16:00).

- All of the respondents were frustrated by text entry:

  How do you feel about entering email on this service?
  I don't like it at all. This was awful. T9 was good for a general message, but I would rather use my keyboard. I don't even like dialing the phone. I usually just use the phone book function on my phone. I like to scroll and select. For reading email, it is good. If I traveled a lot. For composing email it would not be good (R4 27:10).

- Changing modes with softkeys made it difficult for some respondents to enter text.

Navigation

Respondents had some difficulty finding the information they needed. Most of the respondents concentrated on the menus and disregarded other buttons on the telephone. Many of the respondents had difficulty connecting to the Internet, and the same was true for disconnecting.

Illustration C.2 Sprint PCS/NeoPoint 1000 Usability Study: function keys

- The icon for entering the 'Wireless Web' from the phone's main screen was not recognizable (Illustration C.3):
  To make it easier, have an icon or something. Maybe there is; I am not familiar with this phone, so there very well could be a way. I don't know (R4 3:14).
- Most respondents had difficulty connecting to the Internet:
  I know it's not in 'Call History', 'Contacts', 'Schedule', or 'To Do'. It's not 'Sync'. Potentially there would be information about it under 'Preferences', but I would try in either 'MiniBrowser' or 'More'. I'm just going to explore more because I'm curious. (R3 scrolled down to 'More'. R3 pressed LSK, 'Select'. Result: 'More Menu'.) *This is not it* (R3 3:00).
- Some respondents had difficulty ending their connection to the Internet:
  I would go to 'Next' I guess. If it says 'Done', I would press 'Done'. (Task failed.) (R5 58:20).
- Some respondents used the corresponding number on the menu to select menu items:
  I'll press '5'.
  Why would you press '5?'
  Because the number next to it is '5'. (R1 pressed '5'.) *Now it's connected to 'Yahoo!' and I'll look under 'Finance'. So, that's under '5'.* (R1 pressed '5'.) (R1 03:30).
- Some respondents found it easier to scroll and select menu items than to enter text.
- Most respondents were confused by scrolling menu items that were flashing:

Illustration C.3 Sprint PCS/NeoPoint 1000 Usability Study: screen. Reproduced with permission from Sprint PCS

> ***What does it say?***
> *There are zero new messages and there is a test email that is flashing.*
> ***What happened?***
> *When I pressed '2' or 'Check Mail' it brought me back to the same screen* (R2 13:52).

- Selecting incorrect menu items lead to catastrophic errors. Some respondents returned to the same menu time after time, selecting each item when the information they needed was not available from that card:

  > *I'll go to 'Markets'.* (R6 scrolled to highlight 'Markets', then pressed LSK, 'OK'. Result: 'Markets'.) *I'll go to 'Nasdaq'.* (R6 scrolled to highlight 'Nasdaq'. R6 pressed LSK, 'OK'. Result: 'Nasdaq' content card.)
  > ***What came up?***
  > (R6 read aloud from the card.) *I'll go to 'Detailed Info'.* (R6 scrolled to highlight 'Detailed Info', then pressed LSK, 'View'. Result: 'Detail' card.) *No. That's no good.* (R6 pressed RSK, 'DONE'. Result: 'Nasdaq' card. R6 scrolled to highlight '(News)'.)
  > ***What did you push?***
  > *I'm trying to decide if I should go to 'Refresh' or 'News'.*
  > ***What do you think each of them would do?***
  > *I guess 'News' would tell me what is going on in the market. I'll go to that* (R6 09:23).

- Most respondents completed the 'Investor' task efficiently.
- One respondent missed the arrow directing her to scroll (Illustration C.4).
- Today's weather and movie times were unavailable from 'MSN Mobile'.
- One respondent used the 'M' button as a navigation strategy.

Nomenclature

Different services gave various names to menu items. This caused confusion among respondents when they switched from one service to another. Some

6 MSN Mobile
7 Fidelity
8 My Wireless Web
OK ▾ MENU

Illustration C.4 Sprint PCS/NeoPoint 1000 Usability Study: scroll mark

menu items were unrecognizable to the respondents, which caused difficulty with task completion.

- Initial capitals was used by some services while other services used all capitals.
- The labeling of some menu items were misleading to respondents:
  (R5 pressed LSK, 'OK'. Result: 'Yahoo! Finance'. R5 read aloud the menu options.) *'Markets', I wonder if that is stock markets?* (R5 11:50).
- One respondent did not recognize some of the softkey labels as what he needed, which led to task failure:
  Right now it just says 'Next' and 'Alpha'.
  What would you do next?
  I guess 'Enter', but I don't see the word 'Enter'.
  Is there anything that would let you do that?
  I don't know (R6 48:00).

Recommendations

Overall Impressions

- Connecting to the Internet should be made more obvious. Change the icon on the phone's main page to a more intuitive symbol or use the word 'Internet'.
- Respondents requested a personalization feature that would help them navigate to their favorite sites and would allow them to access their email without having to log in each time.
- More instructions should be included about how to use the MiniBrowser with the telephone's instructions. Some respondents stated that they would have read the instructions.

Key Functions

Label keys in an obvious way. Label 'M' as 'Menu', and 'B' as 'Back'.

- Change the color of the 'CLR' key to make it stand out – it's white, like the number keys. Most respondents missed the key, which added to their frustration during task completion.
- The 'SHIFT' key did not meet with respondents' expectations. The 'SHIFT' key function should work as it does on a standard keyboard.
- Softkey labels should be consistent whenever possible. Respondents

reported that keys had too many functions, which led to difficulty using the services.

Text Entry

- Avoid designing sites that contain lots of text entry. All of the respondents found it frustrating to enter text due to the difficulties with triple tap and T9.
- When possible, design cards where users can select items from a list. All of the respondents preferred selecting items from a list to entering text.
- The text entry process should be made faster by speeding up the rate at which users can enter text. This could help users avoid frustration when entering multiple words.
- Make the text entry mode consistent with the type of information the field requires. If a zip code is the required information, the phone should be in numeric entry mode. Conversely, the phone should be in alpha mode for name entry.
- Text entry modes appeared static on the telephone's display. Indicate on the display that text entry modes can be changed.
- Use wizards instead of forms when inputting fewer than three fields of data. Respondents were confused by the forms they encountered during the Movie Task, which contained only two data entry fields.

Navigation

- The Phone.com icon is not a widely recognizable symbol for entering the Internet. Respondents requested an Internet button to make accessing the Internet easier.
- Respondents requested an exit button or closing the keypad cover to end their connection to the Internet.
- Make arrows indicating there is more information below the fold (requiring scrolling) more visible, perhaps blinking them. Higher visibility will increase the likelihood that users will scroll down.
- Content, text entry, and menus are the three types of cards that are currently available on wireless phones. Make an indicator on the screen to orient the user as to what type of card they are viewing.
- Redesign menus to make it more obvious that users can press the number

corresponding to a menu item. This will speed up the process of navigating through menus by alleviating scrolling.

- Avoid having a long list of items on menus. This will help users select the information more easily.
- Design menus that give users a 'way out'. Users need another 'way out' besides the 'Back' button to correct misnavigation.

Nomenclature

- Capitalization should be consistent across cards, services, and platforms. Consistent capitalization will prevent some confusion between menu options and softkey labels.
- Menu items should be common terms that users have encountered on the web.
- The term 'MiniBrowser' is not intuitive. The term 'browser' is generally associated with computers and not phones.

Error Rates

Background

We calculated error rates for each task in order to facilitate our analysis. This section first details how we defined 'steps' and how we calculated respondent errors, and then each task's error rates are provided.

Step Calculation

Steps were counted for each respondent action. 'Steps' were considered to be each action that caused a card to be displayed or refreshed.

- Scrolling was not considered a step.
- Text entry was not considered a step.
- Button presses that resulted in refreshing the current screen were counted as steps. (It was assumed that when a respondent presses a button that they expect some change to occur. The fact that no navigation resulted may have added more to the frustration of the user.)

Correct Steps

(A correct step is a) step that yielded the information desired or resulted in a card that was 'closer' to the information desired.

- Correct steps progressed toward the desired information.
- If the information was directly navigable from a card and the respondent did not select the information, the step was recorded as an error; it did not matter if the information was also navigable from the resulting card.

Error Steps

(An error step is a) step that was not a correct step.

- Steps that resulted in a refresh were always considered errors.
- Any button pressed that did not have a step associated with it was an error. The reasoning was that the respondent intended a step to occur. e.g., hitting the RSK when it was unlabeled, pressing right on the joystick on a nonentry card, etc.
- When respondents used the 'M' button to navigate between cards when the information was available on the displayed card, an error was recorded. Only when it was necessary to change services was 'M' button usage not considered an error.
- Entering incorrect text into a field and moving to the next card was recorded as an error when it the data was 'submitted', not before.

Error Summaries

Activating the MiniBrowser

Activating the MiniBrowser is a fundamental activity that should be obvious to new users of the service. The error rates encountered for this task indicate that MiniBrowser activation is far too difficult and the user interface should be redesigned. (See Table C.1.)

Investor task

The investor task required few steps to successfully complete. The respondents using various services were able to complete this task relatively easily. (See Table C.2.)

Weather task one

Respondents had significant difficulty completing this task. Respondents engaged various services and had difficulty completing the task in a

Table C.1 Success at activating the MiniBrowser

	Success/Fail	Steps Taken	Errors
R1	S	5	4
R2	S	3	1
R3	S	8	6
R4	S	4	2
R5	F	5	5
R6	S	6	5

Note: R1–R6, respondents 1–6; S, success; F, fail.

Table C.2 Success at the investor task

	Success/Fail	Steps Taken	Errors
R1	S	4	0
R2	S	3	0
R3	S	4	0
R4	S	4	0
R5	F	28	17
R6	S	29	14

Note: see Table C.1.

Table C.3 Success at weather task one

	Success/Fail	Steps Taken	Errors
R1	S	12	2
R2	S	16	8
R3	S	22	6
R4	F	18	10
R5	N/A	N/A	N/A
R6	S	10	0

Note: see Table C.1; N/A, not applicable.

reasonable amount of steps. Today's weather is not available from MSN Mobile, which caused respondents to take extra steps. (See Table C.3.)

Weather task two using T9

Not all of the respondents were able to get to this task due to time constraints. The data contained in this table is not convulsive. (See Table C.4.)

Table C.4 Success at weather task two, using T9

	Success/Fail	Steps Taken	Errors
R1	F	N/A	N/A
R2	S	5	0
R3	N/A	N/A	N/A
R4	N/A	N/A	N/A
R5	N/A	N/A	N/A
R6	S	4	0

Note: see Table C.1; N/A, not applicable.

Table C.5 Success at finding movie times

	Success/Fail	Steps Taken	Errors
R1	F	17	3
R2	S	9	1
R3	S	42	15
R4	S	37	26
R5	N/A	N/A	N/A
R6	S	10	0

Note: see Table C.1; N/A, not applicable.

Movie times

This task proved to be the most difficult for the respondents to complete. MSN Mobile did not provide movie times, which caused respondent to take extra steps. (See Table C.5.)

Checking email

Respondents were able to complete the task in a reasonable amount of steps. They reported that checking email was a useful feature. Some respondent reported that personalization would make it easier to complete this operation because it would alleviate the text entry steps. (See Table C.6.)

Sending an email

Respondents were frustrated by this task due to the amount of text entry. Some respondents used the T9 feature of the phone to complete this task with varying degrees of success. (See Table C.7.)

Table C.6 Success at checking email

	Success/Fail	Steps Taken	Errors
R1	S	11	0
R2	S	12	3
R3	S	5	1
R4	S	14	4
R5	F	10	5
R6	S	27	15

Note: see Table C.1

Table C.7 Success at sending an email

	Success/Fail	Steps Taken	Errors
R1	S	8	0
R2	S	10	1
R3	F	26	7
R4	S	7	1
R5	N/A	N/A	N/A
R6	F	6	1

Note: see Table C.1; N/A, not applicable.

Table C.8 Success at exiting the MiniBrowser

	Success/Fail	Steps Taken	Errors
R1	S	6	5
R2	S	1	0
R3	S	7	6
R4	F	12	12
R5	F	1	1
R6	S	1	0

Note: see Table C.1.

Exiting the MiniBrowser

This task was difficult for most respondents to complete. It appeared respondents did not equate hanging up the phone with exiting the Internet. There was an icon for entering the web, but no icon or menu item to exit. (See Table C.8.)

Follow-up Questionnaire

Question 1:
How would you rate the experience of using this phone's MiniBrowser?

Key:

1	2	3	4	5
It was really difficult	I had some difficulty	Neutral	It was enjoyable	It was very enjoyable

The responses are shown in Table C.9

Table C.9 Respondents' experience with using the MiniBrowser

R1	R2	R3	R4	R5	R6
1	3	2	2	1	4

Question 2:
How would you rate the information you found?

Key:

1	2	3	4	5
Not useful at all	Somewhat useful	Neutral	Useful	Very Useful

The responses are shown in Table C.10

Table C.10 Respondents' ratings of information found

R1	R2	R3	R4	R5	R6
3	5	5	4	1	5

Phone Diagram

Features of the phone are shown in Illustration C.5.

Illustration C.5 Sprint PCS/NeoPoint 1000 Usability Study: NeoPoint 1000 phone

Screen Shots

(Screen shots of the main display, phone menu, and MiniBrowser page are shown in Illustration C.6. The wizard and forms architecture are shown in Illustration C.7 and C.8, respectively.)

Illustration C.6 Sprint PCS/NeoPoint 1000 Usability Study: screen shots of main display (left), main phone menu (middle), and main page MiniBrowser (right).

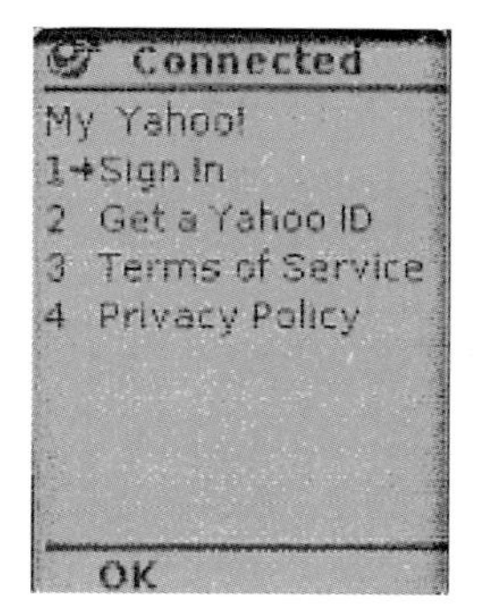

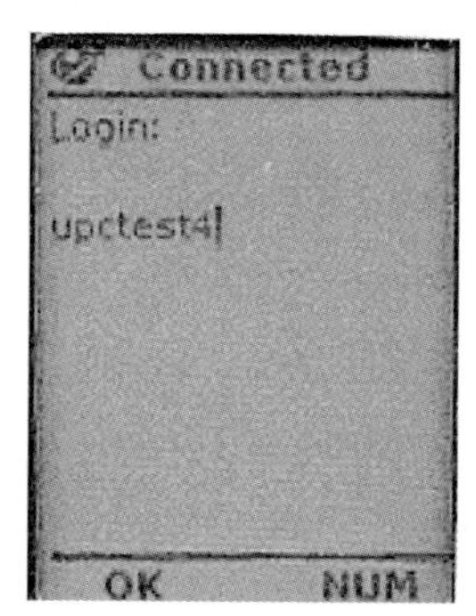

Illustration C.7 Sprint PCS/NeoPoint 1000 Usability Study: screen shots of wizard architecture.

Illustration C.8 Sprint PCS/NeoPoint 1000 Usability Study: screen shots of forms architecture.

Respondent Data

(Respondent data are shown in Table C.11.)
30 November 2000, New York City
All interviews were conducted between 5pm and 8pm, EST.
Each respondent was paid $50 for their time.

Table C.11 Respondent data

Name	Gender	Age [years]	Education	Ethnicity	Own a wireless phone?	Income/year	Web hours/ week
Julie	Female	21–32	College graduate	Asian American	Yes	Student	7–8
Melissa	Female	21–32	College graduate	Caucasian	Yes	55 000–70 000	7–8
Margo	Female	21–32	College graduate	Caucasian	Yes	25 000–40 000	7–8
Ernesto	Male	21–32	College graduate	Asian American	Yes	55 000–70 000	7–8
Allen	Male	33–44	High-school graduate	African American	Yes	25 000–40 000	5–6
Marc	Male	33–44	Some college	Caucasian	Yes	40 000–55 000	2–4

Appendix D
Glossary

Overview

This document was written by Richard Martin and edited by Scott Weiss. It was published through the Wireless Roundtable in 2000. It has been edited significantly and brought up to date for this book.

Definitions

The following are terms and related links that reflect the current language involving 'wireless Web' applications and technologies. The list is by no means exhaustive; it is meant to be sufficient to discuss user interface issues surrounding wireless web technology. All terms are defined with respect to wireless platforms. Terms in bold in the definitions are also defined within this glossary.

1G: first generation of wireless service. This method employed analog technology and was used strictly for voice.

2G (Syn. **PCS**): the current generation of wireless telecommunications including **PCS** in the USA, **GSM** in Europe, and **PDC** in Japan.

2.5G: a packet-switching technology that a majority of carriers will implement in 2001 before continuing on to 3G.

3G: the next generation of wireless telecommunications that is based on new protocols and technologies that will transfer more wireless data at greater speeds and efficiencies. This service will allow for multimedia and database applications not available today. US carriers aren't likely to have infrastructure in place until 2003. Japanese carriers plan to move straight to this technology by 2001.

4G: 'software defined radio'; the next, next thing.

802.11: an IEEE standard for wireless data transfer.

address: **URL**, or **IP** or physical location, or, in Bluetooth™, terminology the 'identity' of a device.

airtime: the time tracked by wireless service providers to determine billing charges. Usage may include sending or receiving calls and other wireless transmissions such as faxes, email, or data files.

alerts: indicators that signal a state change in the device. An incoming message may be tied to a vibration alert while an incoming call is tied to a ringing alert. See also **indicators**.

alias: a label that is used instead of a formal full name. Aliases are commonly used in address books.

alphanumeric: consisting of letters, numbers, and other character types.

analog: a continuous variable signal or circuit or device designed to handle such signals.

antenna: a physical device for sending or receiving radio signals. Antennae come in a variety of shapes and sizes.

API: application program interface.

application: a computer program designed to support the completion of a user task (e.g. Address Book).

application home: the first page of a website or application displayed.

ATM: asynchronous transfer mode; a network technology based on transferring data in cells or packets of a fixed size. ATM supports a range of speeds and can support much higher speeds than Ethernet, the technology used in most **LANs**. ATM creates a fixed channel, or route, between two points whenever data transfer begins. This differs from **TCP/IP**, in which messages are divided into packets and each packet can take a different route from source to destination. This difference makes it easier to track and bill data usage across an ATM network, but it makes it less adaptable to sudden surges in network traffic.

audicon: the audio equivalent of an icon; a brief sound effect.

authentication: the process a wireless transmission network uses to validate a user's identity to prevent unauthorized use.

auto-lock: a feature that locks use of phone when it's turned on. This is disabled by entering an unlocking code via a numeric keypad.

B2B: business-to-business; an e-commerce model whereby businesses transact with other businesses online.

B2C: business-to-consumer; an e-commerce model whereby businesses transact with consumers online.

Back key: used for navigating a wireless device's menu systems and wireless Web cards; similar in function to the Back button on a Web browser.

band: a frequency or contiguous range of frequencies.

bandwidth: the width or capacity of a communications channel. This impacts transmission speeds.

baseband: a Bluetooth™ term referring to a wireless connection between two devices.

base station: a site containing a radio transmitter/receiver and network communication equipment.

beam: the process of transferring data between two devices via **IR** or radio frequency protocols.

beeper (syn. **pager**): a pocket-sized one-way or two-way radio receiver that sounds a tone or vibrates when it receives a transmission. Some pagers currently offered are capable of receiving numeric and alphanumeric messages. Some pagers are also capable of sending messages.

Bluetooth™: a short-range wireless technology designed for local area voice and data communications enabling the exchange of information between many devices. Bluetooth relies on **RF** (radio frequency) communications, and devices must be located within 30 feet of each other. Intel, IBM, Microsoft, Ericsson, Nokia, and Toshiba are participating in the definition of this technology.

bookmark: a stored address that allows the user to access that card's content quickly without the need for entering an **URL**. Methods of creating and accessing bookmarks vary by manufacturer and access provider.

broadband: a term used to compare frequency bandwidth relative to 3 MHz narrowband frequencies. Broadband frequencies can transmit more data at a higher speed than can narrowband frequencies. In general, typical paging services utilize narrowband frequencies. Wireless phones and communication devices use broadband.

browser (syns **web browser, microbrowser, minibrowser**): an application that requests, receives, and interprets information from severs for display to the user.

button [syn. key (hardware)]: (1) in graphical user interface systems the button is a well-defined area within the interface that is clicked to select a command; (2) in hardware, the button is a control with 'on' and 'off' states that is activated by being pressed.

caching: to cache something means to improve the speed of access to it by storing it on the device after it is used the first time. For subsequent uses, the cache-stored data are used, which is significantly faster than accessing the network for the same data. Caches are difficult to manage, since data in a cache can be outdated.

CAD: computer-aided design.

caller ID: caller identification; a feature that displays the name and/or number of the calling party either on a wireless phone screen or a separate device (as is often the case with a landline phone). Virtually all digital phones – as well as many analog phones – have this capability, which may be activated by a wireless service provider.

card (syns **screen**, page): a wireless Web 'page'.

CDMA: code division multiple access; a digital spread-spectrum modulation technique used mainly with personal communication devices such as mobile phones. CDMA digitizes the conversation and tags it with a special frequency code. The data are then scattered across the frequency band in a pseudo-random pattern. The receiving device is instructed to decipher only the data corresponding to a particular code to reconstruct the signal.

CDMA 2000: a higher bandwidth version of **CDMA** backed by Lucent and Qualcomm.

CDPD: cellular digital packet data; a radio technology that supports the transmission of packet data at speeds of up to 19.2 kbps over existing wireless networks. Data are structured into packets, which are transmitted during pauses in wireless telephone conversations, thereby avoiding issues of developing an overlay network for data communications. Packet data over CDPD does not tie up voice channels, which can handle more voice capacity. Packet data also allows AT&T Wireless to charge flat-rate, all-you-can-use pricing because of its spectral efficiency.

cell: the geographic area encompassing the signal range from one base station (a site containing a radio transmitter/receiver and network communication equipment). Wireless transmission networks comprise many overlapping cell sites to use radio spectrum for wireless transmissions efficiently. Also, 'cell' is the basis for the term 'cellular phone'.

cellular: the structure of the wireless transmission networks that are composed of cells or transmission sites. Cellular is also the name of the wireless telephone system originally developed by Bell Laboratories that used low-powered analog radio equipment to transmit within cells. The term '**cellular phone**' is used interchangeably to refer to wireless phones. Within the wireless industry, cellular is also used to refer to non-**PCS** products and services.

cellular phone (syns **handset, mobile phone**): a wireless telephone utilizing **cellular** network technology; see **cellular**.

channel: a specified frequency band for the transmission and reception of electromagnetic signals.

cHTML: compact **HTML**; a variant of **HTML** that is supported by **i-mode** technology.

circuit switching (syn. **session transmission**): transmission method whereby a user is assigned a discrete line or radio channel that is dedicated to the user until the session is complete.

Clear key: a hardware button on most cell phones that erases the character immediately preceding the cursor.

communicator: a mobile telephone with a touch screen and/or a full keyboard. These phones support custom applications, much as **PDAs** do.

configuration: the arrangement of elements of hardware and software both within devices and within networks.

connector: a receptacle or plug for connecting devices to tethers (cables). Many wireless devices have 'syncing cables', which rely on connectors.

context (syn. environment): the environment in which the user is using the device (e.g. 'this application was designed to be used in a driving context').

conversation mode: state in which a cellular phone exists when making or receiving a voice call.

coverage area: the geographic area encompassing a wireless network. This is the area a network service provider offers a cellular service for a subscriber's phone.

CPU: central processing unit; the computer's 'brain'.

cradle: a stand or bracket designed to hold a phone or handheld computer in place on your desktop, or mounted to your automobile dashboard. It may incorporate recharging and/or data-transfer functions.

cursor (syns pointer, highlight): the indicator of active focus on the screen.

customization (syn. **personalization**): the process of changing a device's hardware and software to suit personal needs.

DAP: data access point; a Bluetooth™ term.

data mode: the state in which a phone is accessing the wireless Web whereby it can send and receive data. This state prohibits the answering of incoming calls. Data mode is also the 'wireless Web' access mode.

data services: any service that involves sending or receiving data via a device, including **email**, faxes, **SMS**, and Web access.

dead spot: an area within the coverage area of a wireless network in which there is no coverage or where transmission falls off. Dead spots are often caused by electronic interference or physical barriers such as hills, tunnels, and indoor parking garages.

deck: analogous to a website. A series of connected cards.

device: a network entity that is capable of sending and receiving packets of information and that has a unique device address. These devices range from computer devices (desktop and mobile computers, handheld devices) to phones (cellular, cordless) and office electronics (fax machines, copiers, printers); from computer peripherals (keyboards, headsets, joysticks) to home appliances (toasters, audio equipment), etc.

DHCP: dynamic host configuration protocol (www.dhcp-handbook.com). This protocol enables computers to get their unique Internet addresses when they need them from a server, as opposed to having them assigned more or less permanently.

digital: (1) using a binary code; discrete, noncontinuous values to represent information; analog information can be converted into a digital format; (2) the general term for communications technologies that digitize transmissions into binary code.

display (syns graphic display, **screen, LCD screen**): the screen on a wireless device. Defined by 'color' or 'black and white', pixel dimensions, number of lines, and whether backlit or not.

domain: (1) a task area (e.g. the 'Create List' domain); (2) 'foo' in www.foo.com.

DSSS: direct sequence spread spectrum; a radio transmission technology designed to reduce a signal's susceptibility to interference (www.webopedia.com).

DTD: document type definition: a language that describes the contents of an **SGML** document. The DTD is also used with **XML**, and the DTD definitions may be embedded within an XML document or in a separate file. DTDs are expected to be replaced by an XML schema from the **W3C**.

DTMF: dual tone multifrequency; tones that communicate with computerized systems, such as answering machines, voice mail, or when banking by phone.

dual band: a wireless phone that is capable of operating on two frequency bands such as the 800 MHz digital band and the 1900 MHz digital **PCS** band.

duplex/full duplex: simultaneous two-way transmission, such as experienced in a phone conversation. In contrast, many speakerphones are half-duplex and will transmit in only one direction – from the loudest noise – at a time.

EDGE: enhanced data for **GSM** evolution; a third-generation (**3G**) wireless technology standard approved by the International Telecommunications Union (**ITU**). EDGE converges GSM (global system for mobile communications) and **TDMA** (time division multiple access) technologies. AT&T Wireless currently uses TDMA IS-136 technology. The combined systems will enable people to use their EDGE devices in most major cities around the world. EDGE is intended to support transmission rates of up to 384 kbps.

EDI: equipment Identifier for a data device.

email: electronic mail.

email ID (syn. **user ID**): a short sequence of numbers and/or letters used to identify a specific user on an email server; most times used in conjunction with a password or **PIN**.

encryption: a procedure used to convert data to code in order to prevent any but the intended recipient from reading it.

End key: the physical button that disconnects calls.

EPOC (syn. **Symbian OS**): Symbian's technology for wireless information devices that includes an operating system and provisions for applications, connectivity, and software development.

ETSI: the European Telecommunications Standardization Institute, a not-for-profit organization whose mission is to produce telecommunication standards in Europe.

express key (syns joystick, **navigation button, navigator key**): navigation joystick control with up, down, left, and right directionals. Additional functions are tied to several of these directionals if the key is held for a longer time.

FCC: Federal Communications Commission, USA; formerly the **FRC**.

field: an element of a database record in which one piece of information is stored.

FRC: Federal Radio Commission, USA; now the **FCC**.

FTP: file transfer protocol.

gateway (syns **router**, proxy server): software or hardware that enables communication between computer networks that use different communication protocols (see **router**).

global navigation: navigation at the top level of a deck.

GPRS: general packet radio service; a mobile data communication service that is widely expected to be the next major step in the evolution of **TDMA** and **GSM**. It will be a packet-based system that could be used for 'bursty' data applications such as mobile Internet browsing, **email**, and push technologies. GPRS has been demonstrated as fast as 115 kbps, but may be initially deployed with speeds of 28.8 kbps.

GPS: Global Positioning System; satellite technology that is able to pinpoint exactly the location of a GPS receiver.

Graffiti® software: an input method for characters, utilizing a **stylus.**

graphic display: see **display.**

graphics: visual representations such as pictures, charts, and icons.

GSM: Global System for Mobile communication; the de facto digital cellular radio network in Europe (includes GSM 900, GSM 1800, and GSM 1900); a subset of **PCS.**

GUI: graphical user interface; a GUI (usually pronounced goo-ee) is a graphical rather than a textual user interface to a computer. The term came into existence because the first interactive user interfaces to computers were text and keyboard oriented and usually consisted of command strings.

handheld computer (syn. **PDA**): a computer that can conveniently be stored in a pocket (of sufficient size) and that can be used while you're holding it. Some handheld computers accept handwriting as input; others have keyboards.

hands-free operation/handling: using a wireless phone without having to lift or hold the phone to your ear.

handset (syn. **cellular phone; mobile phone**): a wireless telephone utilizing cellular network technology; see **cellular.**

HDML: handheld device markup language; forerunner of **WML**. This is the predominant language used for the wireless Web in the USA. WML sites are translated to HDML when passing through the **gateway.**

HDTV: high-definition television; a digital video format with significantly higher resolution than existing television formats.

headline (syns heading, header, title): emboldened title that appears at the top of every card. Used primarily as a navigation aid.

help (syn. online help): information provided to facilitate users' ability to find information and perform transactions. Help is frequently accessed through a 'Help' menu item or **softkey.**

highlight (syns **cursor, pointer**): the content area of the screen which contains the focus. Each device has proprietary highlighting; some employ inverse pixels, some employ a graphic, some use brackets ('[').

hits: number of items that match a set of search criteria.

home: the default site that is displayed when the **browser** is turned on.

Hotkey (syns one-touch dialing, speed dialing): the button one can press to dial a stored phone number.

HotSync® operation (syns **sync, synchronize**): an operation used to synchronize the contents of a **PIM** (or a device working in that capacity) with another device (typically a computer, but may be other information sources; i.e. Web mail).

HTML: hypertext markup language; a markup language used to structure text and multimedia documents and to set up hyperlinks between documents; used extensively on the World Wide Web. See also **SGML** and **XML**.

HTTP: hypertext transfer protocol; a protocol used to request and transmit files, especially Web pages and Web page components over the Internet or other computerized networks.

hyperlink: see **link**.

iDen: integrated digital enhancement network; a Motorola-developed air interface standard used by Nextel in its SMR (specialized mobile radio) based mobile network for voice and data services. The standard is based on the **GSM** standard but allows for two-way radio capabilities as well.

JEEE: Institute of Electrical and Electronic Engineers, USA.

IMAP (Syns **POP3**, **SMTP**): Internet message access protocol, version 4.

IMEI: the international mobile equipment identifier; a number that uniquely identifies an individual wireless phone or communicator. The IMEI appears on the label located on the back of the phone and is automatically transmitted by the phone when the network asks for it. A network operator might request the IMEI to determine if a device is in disrepair, stolen, or to gather statistics on fraud or faults.

i-mode: packet-based information service for mobile phones from **NTT** DoCoMo (Japan), first to provide Web browsing from cellular phones.

inbox: area where new incoming messages (voice or electronic) are stored.

incoming call: a call placed by another party to your wireless phone. In virtually all current wireless phone service plans in the USA, the owner of the wireless phone pays for all calls, both incoming (calls from others) and outgoing (placed by you).

indicators: audio, vibration, or visual (icons or light) cues that alert the user to the state of the device. See also **alerts**.

input field: area in which to enter information.

input form: a card with multiple input fields.

Internet capability: a wireless communication device that is capable of Internet functions such as **email** (send, receive, forward, or reply) and browsing the World Wide Web.

interoperability: the ability of software and hardware on multiple machines from multiple vendors to communicate.

IP: Internet protocol; the most important of protocols on which the Internet is based. IP is a standard describing software that keeps track of the Internet's addresses for different nodes, routes outgoing messages, and recognizes incoming messages. It allows a packet to traverse multiple networks on the way to its final destination.

IP address: Internet protocol address; the address of a computer attached to a **TCP/IP** network. Every client and server station must have a unique **IP** address. Client workstations have either a permanent address or one that is dynamically assigned to them each dial-up session. IP addresses are written as four sets of numbers separated by periods; for example, 204.171.64.2.

IR: infrared; IR communications use light that is invisible to the human eye and require a clear 'line of sight' between devices.

IrDA: (1) Infrared Data Association (www.irda.com); an international organization that creates and promotes interoperable, low-cost infrared (**IR**) data interconnection standards that support a walk-up, point-to-point user model; (2) a protocol suite designed to support transmission of data between two devices over short-range point-to-point infrared at speeds between 9.6 kbps and 4 Mbps; (3) that small semitransparent red window that you may have wondered about on your notebook computer.

IR port (syn. IR data port): infrared (**IR**) communication port, used to '**beam**' information to and from other IR ports.

ISO: International Standardization Organization.

ITU: International Telecommunications Union.

key (syn. **button**): a hardware control that when pressed sends a character code to the monitoring software.

keypad lock: allows the mobile phone user to disable the keys so that a number will not accidentally be dialed while the phone is in a pocket or purse, for example.

LAN: local area network.

LCD screen: liquid crystal display screen; these have low energy requirements and are generally easy to read. LCD screens are made by sealing a liquid compound between two pieces of glass and/or a filter. The screen has hundreds or thousands of dots that are charged or not charged, making them reflect or not reflect light to form letters, characters, and numbers. Some LCD screens have an electroluminescent panel behind them and are said to be 'backlit'.

LDAP: lightweight directory access protocol; a protocol used to access a directory listing. LDAP support is being implemented in Web browsers and **email** programs, which can query an LDAP-compliant directory. It is expected that LDAP will provide a common method for searching email addresses on the Internet, eventually leading to a global 'white pages'. LDAP is a sibling protocol to **HTTP** and **FTP** and uses the 'ldap://' prefix in its **URL**.

LED: light-emitting diode; they can be red, amber, green, or blue, and are used in small displays. They are used as indicator lights on wireless devices when **LCD screens** are not appropriate.

link (syn. **hyperlink**): a reference (link) from some point in one hypertext document to (some point in) another document or another place in the same document. A browser usually displays a hyperlink in some distinguishing way (e.g. in a different color, font, or style). When the user activates the link (e.g. by clicking on it with the mouse) the browser will display the target of the link.

local service area: the geographic area that telephones may call without incurring roaming or long-distance charges.

log on (syn. login): the process of entering **user ID** and **password** information.

login procedure: the process of entering **user ID** information and **password** authentication.

MAN: multiple activation number; a number assigned to a wireless device that functions instead of a modem **IP** address.

mark: the process of creating a bookmark. Some phones have a 'Mark' **key**.

MCDN: micro cellular digital network; a foundation of the wireless architecture that Ricochet (a radio transmission data network) uses. It is composed of four primary technology elements: the 902–928 MHz frequency band, spread-spectrum/frequency-hopping, packet-switched networking, and protocols for transmitting and receiving data.

Menu Help: Help screens accessed through pressing the Menu button.

Menu key: a **key** that is pressed to get to a phone-based menu. Not present on all phones.

menu: list of options or choices that the user may pick from.

metalanguage (syn. **XML**): a language used to define other languages. With respect to **XML**, it represents the contextual meaning of the data in order to facilitate the information-sharing process.

messaging: using various products, services, and technologies to transfer messages from one person to another, or from one device to another such as traditional numeric or **alphanumeric** paging, **email** or short messages (see **SMS**) delivered to wireless devices.

microbrowser: a Web browser specifically designed to run on **PDAs** and cellular telephones (see **cellular phone**).

minibrowser: see **microbrowser**.

MMS: multimedia messaging service.

mobile phone (syns **cellular phone, handset**): a term often used interchangeably with **cellular phone** or wireless phone. A mobile phone is distinguished from a 'cordless phone', by its significantly longer range.

Mobitex: a packetized wireless 900-MHz wide area network (**WAN**) that allows mobile or portable subscribers to transfer data, including **email**, through the growing national and international network infrastructure.

mode: the state of a device or an application. Depending on the mode, **softkeys** and other **buttons** have different functions.

modem: abbreviation of modulator/demodulator; the modem converts digital computer signals into analog form for transmission over analog telephone systems.

MR: memory recall.

multi-tap: method for entering **alphanumeric** text by taping **buttons** multiple times for each letter.

My Menu List: personalized portal of **NTT** DoCoMo's recommended sites. Only NTT DoCoMo's sponsored sites maybe added to this list.

narrow band: mobile and portable radio (including paging) services such as two-way paging, acknowledgement paging, voice paging, and data services. These services are transmitted over a set of frequencies set aside by the **FCC** in 1994.

navigation button (syns joystick, **navigator key, express key**): up, down, left, and right directionals. Additional functions are tied to several of these directionals if the key is held for longer times.

navigation icons: images that appear on a cellular phone's display that indicate where the user may scroll to view more of the card or menu. See also **scroll bar**.

navigator key (syns joystick **navigation button**, **express key**): a joystick device for moving the cursor around the display.

NCSA: National Center for Supercomputing Applications, USA.

NEI (syn. **IP address**): network entity identifier.

network: the infrastructure enabling the transmission of wireless signals. A network ties things together and enables resource sharing.

NSF: National Science Foundation, USA.

NTSC: National Television Standards Committee. The NTSC format is a television display format used in many countries. It is the oldest existing standard and was developed in the USA. It was first used in 1954 and consists of 525 horizontal lines of display and 60 vertical lines. It is sometimes referred to as Never Twice The Same Color (http://kropla.com/tv.htm).

NTT: Nippon Telephone and Telgraph.

numeric: most often refers to messaging services that are capable of transmitting numbers only (no letters), or a wireless message that contains only numbers such as the phone number page on a numeric pager.

numeric page: where callers enter numbers in the voicemail box instead of a voice message.

Operating system: see **OS**.

OS: operating system; the program that, after being initially loaded into the computer by a 'bootstrap program', manages all the other programs in a computer. The other programs are called **applications**. The applications make use of the operating system by making requests for services through a defined application program interface (**API**). In addition, users can interact directly with the operating system through an interface such as a command language.

packet: a component of data. Data streams are broken into packets.

packet switching (syn. packet transmission): a transmission method where data packets from a number of different conversations or data messages can transverse the same channel. Packets are mixed on the channel but are reassembled correctly at the receiving end.

pager (syn. **beeper**): a pocket-sized one-way or two-way radio receiver that sounds a tone or vibrates when it receives a transmission. Some pagers currently offered are capable of receiving numeric and **alphanumeric** messages. Some pagers are also capable of sending messages.

pairing: developing a trusted relationship between two devices.

Palm OS®: Palm Operating System; the operating system software upon which the Palm, Handspring, and other devices are based.

PAL: phase alternating line; developed in the United Kingdom and Germany. It was first used in 1967 and is a 625 × 50 line display. Proponents call it Perfection At Last. Owing to the cost of the enormous circuit complexity, critics often refer to it as Pay A Lot. Different types use different video bandwidth and audio carrier specifications.

Common types are B, G, and H; less common types include D, I, K, N, and M. The different types are generally not compatible (http://kropla.com/tv.htm).

PAN: personal area network; allows devices to work together and share information and services. Using **Bluetooth™** wireless technology, PANs can be created in public places, in your home, in your office, and even in your car. This network enables everyday devices to become smart, tetherless, devices – working and communicating together. For example, it offers the ability to 'wirelessly' synchronize with your desktop to access your **email** and Internet or intranet from remote locations.

password (syn. **PIN**): a secret number used to identify users and devices.

pauses: special characters inserted in the **DTMF** that produce a delay in the dial tone sequence.

PC card: personal computer card; see **PCMCIA card.**

PCMCIA: Personal Computer Memory Card International Association; a group of hardware manufacturers and vendors responsible for developing standards for **PC cards** (or **PCMCIA cards**).

PCMCIA card (syn. **PC card**): removable, credit-card-sized devices that may be plugged into slots in PCs (personal computers) and wireless communication devices to provide fax or modem functions or network cards.

PCS (syn. **2G**): personal communications service or system; generally, a marketing term used to describe a wide variety of two-way digital wireless service offerings operating at 1900 MHz. PCS services include next-generation wireless phone and communication services, wireless local loop, inexpensive walk-around communication services, with lightweight, low-powered handsets, in-building cordless voice services for business, in-building wireless **LAN** services for business, enhanced paging services, as well as wireless services integrated with wired networks. A PCS refers to the hardware and software that provide communication services.

Inclusive of CDMA, TDMA, and GSM technologies.

PDA (syn. **PIM**): personal data assistant; a handheld device acting as an electronic organizer.

PDC: personal digital communications; a packet-switching technology used in Japan.

personalization (syn. **customization**): the ability to customize an application or hardware device to one's needs.

phone book (syns address book, contacts): the collection of telephone numbers the user has stored on a device.

picocell: a wireless base station with extremely low output power, designed to cover an extremely small area, such as one floor of an office building.

piconet (syn. **PAN**): at least two devices communicating together, allowing up to eight devices.

PIM (syn. **PDA**): personal information manager; e.g., handheld devices acting as electronic organizers or calendaring programs such as Microsoft Outlook.

PIN (syn. **password**): personal identification number: a secret number used in identifying users and devices.

platform: with reference to computers, a platform is an underlying computer system on which application programs can run. On personal computers, Windows 95 and the Macintosh are examples of two different platforms. On enterprise servers or mainframes, IBM's System/390 is an example of a platform.

pointer: see **cursor.**

POP3 (syns **IMAP, SMTP**): version 3 of the Post Office Protocol, the basis for the global **email** standard. It was written in November 1988 by Marshall Rose. POP3 allows a user to read email from an email server via a **TCP/IP** or other type of connection. Sending mail requires **SMTP** or some other method. See **SMTP.**

port: input–output connection of a device that is used to communicate with other devices.

portal: a website or service that provides services (news, weather, sports scores) and links to other sites.

POTS (syn. **PSTN**): the 'plain old telephone service'. See **PSTN.**

Proxy server: see **router** and **gateway.**

PSTN (syn. **POTS**): public switched telephone network; the world's collection of interconnected voice-oriented public telephone networks, both commercial and government-owned.

pull: downloading information from a server; browsing the Web is a example of the 'pull' model.

push: a data distribution technology in which selected data are automatically delivered into the user's computer at prescribed intervals or based on some event that occurs.

QWERTY: the standard arrangement of a US or UK keyboard, named for the first six letters on the second row of the keyboard.

RAM: random access memory.

range: the distance that a wireless device can be located from a **repeater** and still receive and send data reliably.

repeater: receives radio signals from the **base station,** which are then amplified and retransmitted to areas where radio shadow occurs. Repeaters also work in the opposite direction; i.e., receiving radio signals from mobile telephones, then amplifying and retransmitting them to the base station.

RF: radio frequency.

roaming: using a wireless phone outside of your service provider's local coverage area or home calling area. Roaming arrangements between service providers expand the potential area for phone use. Service providers typically charge a higher per-minute fee for calls placed outside their home calling or coverage area.

ROM: read only memory.

router (syns **gateway**, proxy server): a device that forwards data packets from one local area network (**LAN**) or wide area network (**WAN**) to another. Based on routing tables and routing protocols, routers read the network address in each transmitted frame and decide how to send it based on the most expedient route (traffic load, line costs, speed, bad lines, and so on).

S-CDMA: synchronous code division multiple access; a proprietary version of code division multiple access (**CDMA**), S-CDMA was developed by Terayon Corporation for data transmission across coaxial cable networks. S-CDMA scatters digital data up and down a wide frequency band and allows multiple subscribers connected to the network to transmit and receive concurrently. This method of data transmission was developed to be secure and extremely resistant to noise.

Scratch Pad: an application that allows user to store numbers on the phone during conversations.

screen (syns **card**, page): (1) see **card**; (2) the **LCD** display of a cellular phone.

screen saver: a mode that deactivates the **cellular phone** screen after a set amount of idle time to conserve battery power.

scroll bar: a scroll bar appears on the right of the screen on Nokia phones indicating the position in a menu structure. 'Tabs' in this context represent menu items. See also **navigation icon**.

scroll button (syn. **navigation button**): used to navigate through phone menus and submenus. Usually denoted by up–down arrows.

search engine: software that finds data based on criteria entered by the user.

section navigation: user movement within the domain of a particular service or Web deck.

security code: a number used to prevent unauthorized or accidental alteration of data programmed into wireless phones. The security code can be used by the owner of a phone to change the lock code.

Send key (syn. **Talk key**): dials the displayed number or answers the phone.

server: a computer in a network shared by multiple users. The term may refer both to the hardware and to the software or just the software that performs the service. For example, the term 'Web server' may refer to the Web server software in a computer that also runs other applications or it may refer to a computer system dedicated only to the Web server application. There would be several dedicated Web servers in a large website.

service (syn. site): (1) a site that provides the ability to accomplish tasks (to book plane tickets, view TV programming schedules, etc.); (2) applications running over a wireless connection.

service agreement: a business contract or agreement that outlines the services provided and the costs for the services by a wireless service provider. Service agreements typically include a monthly base rate (with included minutes) and per-minute charges for minutes over the monthly maximum.

service links: links provided by a service, such as a **portal.**

session transmission (syn. **circuit switching**): a transmission method that assigns users a discrete line or radio channel that is dedicated to the users until the session is complete.

SGML: standard generalized markup language; an **ISO** standard for defining the format in a text document. An SGML document uses a separate document type definition (**DTD**) file that defines the format codes, or tags, embedded within it. Since SGML describes its own formatting, it is known as a metalanguage. SGML is a very comprehensive language that includes hypertext links. The **HTML** format used on the Web is an SGML document that uses a fixed set of tags. See also **HTML** and **XML.**

SIG: special interest group.

SIM: subscriber identity module; a card commonly used in a **GSM** phone. The card holds a microchip that stores information and encrypts voice and data transmissions. The SIM card also stores data that identifies the caller to the network service provider.

smart phone: a communicator; a handheld device that features **PDA** and telephone functionality; they look and behave more like PDAs or pagers than they do telephones, but support both. See also **communicator.**

smart punctuation: T9 system through which punctuation is added; see **T9 text input.**

SMR: specialized mobile radio; see **iDen.**

SMS: short message service; a service for sending messages of up to 160 characters to mobile phones that use the Global System for Mobile (**GSM**) communication. GSM and SMS services are primarily available in Europe. SMS is similar to paging; however, SMS messages do not require the mobile phone to be active and within range, and they will be held for a number of days until the phone is active and within range. SMS messages are transmitted within the same **cell** or to anyone with roaming capability. They can also be sent to digital phones from a carrier's website or from one digital phone to another.

SMTP (syn. **POP3**): simple mail transfer protocol; this server-to-server protocol is used to transfer **email** messages. POP3 is used to retrieve messages. See **POP3**.

softkey: hardware buttons on a **cellular phone** the function of which changes depending on the **card** or **menu** currently displayed. The function is indicated by text that appears on the display above the **button**.

standby mode: the state when the phone is on and registered on a network but not making or receiving calls.

start screen: what is displayed on the **screen** when the telephone is turned on.

status indicators: Icons that display information about the status of the phone and/or phone service or system state.

Stinger: a mobile phone technology standard based on Windows CE. Acknowledging that there was no such thing as an ideal mobile device, Microsoft will support three different technologies: the already announced Pocket PC platform for handheld devices, a 'feature' phone, and a smart phone (see **smart phones**).

stylus: a pen-like device for character input on a handheld device.

Symbian OS (syn. EPOC): Symbian's technology for wireless information devices that includes an operating system, and provisions for applications, connectivity, and software development.

sync (**HotSync® operation, synchronization**): an operation used to synchronize the contents of a **PIM** (or a device working in that capacity) with another device (typically a computer, but may be other information sources, i.e. Web mail).

synchronization (syns **sync, HotSync® operation**): see **sync**.

T9 text input: software that reduces the number of key presses required to enter text with a numeric keypad using a database of commonly used words. This technology was developed by Tegic, which was later acquired by AOL Mobile.

Talk key (syn. **Send key**): initiates the dialing sequence for a cellular telephone call.

task: (1) a step in a process; (2) a function to be performed.

TCP/IP: transfer control protocol/Internet protocol.

TDMA: time division multiple access; a method of digital wireless communications transmission allowing a large number of users to access a single radio-frequency (**RF**) channel without interference. Each user is given a unique time-slot within each channel.

telephony: communication, often two-way, by means of electrical signals carried by wires or radio waves.

text entering mode: different modes allow for different methods of text entry; modes include T9, Alpha, Num, and Sym.

text message: **email** addressed to a **cellular phone**.

Times Square Scrolling: when a mobile telephone handset displaying WAP content, blink-scrolls the text. This only occurs when the text is longer than a single line can display, and the code stipulates that it will not 'wrap'.

TLS: transport layer security; a **GSM** term.

trackwheel [syn. Navi Roller™ (Nokia)]: rolling control that is used to scroll or move a pointer or cursor vertically.

triple tap: the act of having to press a telephone keypad key up to three, or even more, times to produce the desired character; for example, the letter 'c' is produced by pressing the '2' key three times.

up.browser: OpenWave's **browser** technology specifically designed to work within the limited resources of a **cellular phone**.

UMTS: universal mobile telecommunications system; the technology envisioned for the next generation of **GSM** (Global System for Mobile communications). UMTS is a wireless standard approved by the International Telecommunications Union (**ITU**) and is intended for advanced wireless communications. UMTS will provide extremely-high-speed mobile data, advanced multimedia capabilities, and serve as a platform for other advanced wireless products and services.

URL: universal resource locator; the global address of documents and other resources on the World Wide Web. The first part of the address indicates what protocol to use, and the second part specifies the **IP address** or the **domain** name where the resource is located.

user ID: user identification; a short sequence of numbers and letters used to identify a specific user on a network or device; most times used in conjunction with a **password** or **PIN**.

voice interface: preprogrammed voice command capabilities.

voicemail (syn. voice message): a voice message recorded by phone and stored in a 'mail box' for later review.

W3C: World Wide Web Consortium; a standards body that develops protocols to promote interoperability, including standard languages such as **HTML** and **XML**.

WAN: wide area network.

WAP: wireless application protocol: a protocol developed to allow intelligent transmission of optimized Internet content to wireless phones.

wCDMA: wideband code division multiple access; higher **bandwidth** version of **CDMA** backed by Ericsson, Nokia, and Japanese handset manufacturers.

web browser: see **browser**.

Web clipping: a format for delivery of web-based information to handheld devices and **cellular phones**. Websites are truncated or otherwise automatically edited to fit on small screens.

Windows CE: an operating system based on the Microsoft Windows OS but designed for including or embedding in mobile and other space-constrained devices. Although Microsoft does not explain the 'CE', it is reported to have originally stood for 'Consumer Electronics'.

wireless carrier: a company that provides wireless telecommunication services.

wireless network: network connectivity without a tether.

wizard: a means of collecting information from the user, typically one field at a time.

WLAN: wireless local area network.

WML: wireless markup language; a tag-based display language providing navigational support, data input, hyperlinks, text and image presentation, and forms. A browsing language similar to Internet **HTML**. This standard is not yet in wide use in the USA. Sites written in WML are translated to **HDML** when passing through the **gateway** to the requesting device.

WTA: wireless telephony application.

XML: extensible markup language; a flexible way to create common information formats and share both the format and the data on the World Wide Web, intranets, and elsewhere. XML is currently a formal recommendation from the World Wide Web Consortium (**W3C**). XML is similar to the language of today's Web pages, **HTML**. XML is 'extensible' because, unlike HTML, the markup symbols are unlimited and self-defining. XML is actually a simpler and easier-to-use subset of the Standard Generalized Markup Language (**SGML**), the standard for how to create a document structure. It is expected that HTML and XML will be used together in many Web applications. See also **HTML**, **metalanguage**, and **SGML**.

Sources

We referred to the following sources for this information (all on the World Wide Web):

AnyWhereYouGo.com
ATT.com (AT&T)
Bluetooth.com
cnet.com/Resources/Info/Glossary
Dictionary.com
Ericsson.com
Microsoft.com
Motorola.com
NTTDoCoMo.com
Palm.com
OpenWave.com
Symbian.com
Telenor.com
WAPForum.com
WirelessData.org

and to

Industry Standard Grok Issue on 'Wireless', November 2000, out of print.

Mitsubishi T250 Users' Manual,
www.mitsubishiwireless.com/manuals/index.html.

Sprint NP1000 Users' Manual (same as the user manual for the NeoPoint 1000), out of print.

Bibliography

This bibliography includes references that were cited or used to provide the content in this book, as well as additional readings. Most of the research for this book was conducted on the Web and, for your convenience, websites have been separated from books and journals.

Books and Journals

Apple Computer Inc., 1987, *Human Interface Guidelines: The Apple Desktop Interface*. Addison Wesley, Reading, Massachusetts.

Apple Computer Inc., 1996, *Newton 2.0 User Interface Guidelines*, Addison Wesley, Reading, Massachusetts.

Ballard, Barbara and Miller, Bob, 2000, *Sprint HDML Style Guide Version 3.0*, Product Design and Usability, Sprint PCS, October 2000.

Bates, Bob and Lomothe, Andre, 2001, *Game Design: The Art & Business of Creating Games*, Premier Press, Indianapolis, IN.

Bergman, Eric, (ed.), 2000, *Information Appliances and Beyond*, Morgan Kaufman, San Francisco.

Bias, Randolph and Mayhew, Deborah, 1994, *Cost-justifying Usability*, Morgan Kaufmann, San Diego, CA.

GSM Association, 2001, *M-Services Guidelines*, 31 May 2001.

Jerkins, Andrejs and Todorov, Ivaylo N, 2001, *Critical Success Factors in the New Economy: How to Make WAP Worth IT; A Study of Wireless Internet*, masters thesis in Service Management, Spring 2001, Department of Business and Economics, Service Research Centre, Karlstad.

Kaplan, Jerry, 1994, *Startup*, Penguin Books, New York.

Kounalakis, Markos, 1995, *Defying Gravity: The Making of Newton*, Beyond Words Publishing, Hillsboro, Oregon.

Kovitz, Benjamin, 1999, *Practical Software Requirements*, Manning, Greenwich, CT.

Krug, Steve and Black, Roger, 2000, *Don't Make Me Think! A Common Sense Approach to Web Usability*, New Riders Publishing, Indianapolis, IN.

Moore, Geoffrey A. and McKenna, Regis, 1999, *Crossing the Chasm: Marketing and Selling High-tech Products to Mainstream Customers*, Harperbusiness, New York.

Pearrow, Mark. *The Wireless Web Site Usability Handbook*, 2002, Charles River Media, Hingham, MA.

Phone.com, 2000, *HDML Application Style Guide*, Openwave, Redwood City, CA.

Robertson, Suzanne and Robertson, James, 2000, *Mastering the Requirements Process*, Addison Wesley, Reading, MA.

Rouse, Richard, Ogden, Steve (Illustrator) and Rybczyk, Mark Louis, 2001, *Game Design: Theory and Practice*, Wordware Publishing. Plano, TX.

Rubin, Jeffrey, 1994, *Handbook of Usability Testing*, John Wiley, Chichester, UK.

Smith, Douglas K. and Alexander, Robert C., 1999, *Fumbling the Future: How Xerox Invented, Then Ignored, the First Personal Computer*, iUniverse.com.

WAP Forum, 2000, *Wireless Application Protocol White Paper, Wireless Internet Today*, WAP Forum, June 2000, www.wapforum.com/what/WAPWhite_Paper 1.pdf.

Wurman, Richard Saul, 1995, *Information Architects*, Watson-Guptill Publications, New York.

Websites

Asahina, Jon, 2000, 'A Simple Comparison of Bluetooth TM and 802.11b; A TROY XCD White Paper', Troy XCD, www.troyxcd.com.

Barnett, Shawn, 2000, 'Jeff Hawkins: The Man Who Almost Single-handedly Revived the Handheld Computer Industry', *Pen Computing Magazine*, www.pencomputing.com/palm/Pen33/hawkins1.html.

BBC News, 2001, 'A history of Psion', 11 July 2001, news.bbc.co.uk/hi/english/ business/ newsid_1434000/1434369.stm.

CIC Corporation, 2001, 'CIC Jot®', November 2001, Redwood Shores, CA, www.cic.com.

Eurotechnology Japan K. K., 2001, 'The Unofficial Independent imode FAQ', Tokyo, Japan, www.eurotechnology.com/imode/faq-gen.html.

Farley, Tom, accessed 2002, 'Mobile Telephone History', TelecomWriting.com, www.privateline.com/PCS/history.htm.

Giangrasso, Dom with Blickenstorfer, Conrad H., 1996, 'My Love Affair With Pen Computers', *Pen Computing Magazine* number 5, April 1996, www.pencomputing.com/TabletPC/pen_history_dom.html.

Gilbert, Howard, 1995, 'Introduction to TCP/IP', *PCLT (PC Lube and Tune)*, February 1995, www.yale.edu/pclt/COMM/TCPIP.HTM.

Harwell, *et al.*, 1993, 'What is a Requirement?', in *Proceedings of the Third International Symposium of the NCOSE*, 1993, www.incose.org/rwg/ what_is.html.

Hendrick, Hal W., 1996, 'Good Ergonomics is Good Economics', HFES Presidential Address, 40th Annual Meeting, Santa Monica, CA, hfes.org/publications/goodergo.pdf.

Howe, Denis, 1993, 'Free On-line Dictionary of Computing', www.foldoc.org.

Hutton, Ian, 2001, 'Implementing Applications The Quartz Way', Symbian Ltd. August 2001, www.symbiandevnet.com/techlibrary/papers/qapps/ianh.html.

KC, 2002, 'Computer Closet Home Page', January, 2002, www.geocities.com/~compcloset/index.html.

Meyer, André, 1995, 'Pen Computing', *ACM SIGCHI Bulletin*, July 1995, www.amug.org/amug/sigs/newton/nanug/PenReport/NewPenCom.html.

Microsoft Corporation, 2000, 'Designing a User Interface for Windows CE', www.msdn.microsoft.com/library/default.asp?url=/library/en-us/wcedesgn/htm/uidsn.asp.

Motorola Inc., 2001a, 'Motorola History', www.motorola.com/General/Timeline/.

Motorola Inc., 2001b, 'Timeport P935 SDK On-line Documentation', www.developers.motorola.com/developers/wireless/downloads/p935_download.html.

Nokia, 2000, 'The User Interface of the Nokia 9210 Communicator', Nokia Mobile Phones Customer Application Development Group, http://download.forum.nokia.com/download/Nokia_9210_style_guide.pdf.

Nokia Corporation Mobile Software Unit, 2001, 'Designing Applications for Smartphones – Series 60 Platform Overview', 16 November 2001, http://download.forum.nokia.com/download/Series60-desgningapplications.pdf.

Pen Computing Magazine, 2001, 'Magic Cap/DataRover Online', November 2001, www.pencomputing.com/magic_cap/.

Polsson, Ken, 2001, 'Chronology of Handheld Computers', www.islandnet.com/~kpolsson/handheld/.

RIM, accessed 2002, 'Developer's Guide: RIM 950 Wireless Handheld™ User Interface Engine API Version 1.7', www.blackberry.net/developers/na/c_plus/knowledge/documentation/download/ui_20_datatac.pdf.

Sprint PCS Application Developer's Program, 2000, 'HDML Flowchart Template', www.developer.sprintpcs.com/downloads/index.cfm.

Trochim, Bill, accessed 2002, 'Center for Social Research Methods', www.trochim.human.cornell.edu.

Index